Sustainable Development in Southern Europe

Eric Vaz • Teresa de Noronha

Sustainable Development in Southern Europe

Spatial Analysis of Regional Challenges

Springer

Eric Vaz
Department of Geography
and Environmental Studies
Ryerson University
Toronto, ON, Canada

Teresa de Noronha
Faculty of Economics
University of Algarve
Faro, Portugal

ISBN 978-3-662-62175-2 ISBN 978-3-662-62177-6 (eBook)
https://doi.org/10.1007/978-3-662-62177-6

This Springer imprint is published by the registered company Springer-Verlag GmbH, DE part of Springer Nature.
The registered company address is: Heidelberger Platz 3, 14197 Berlin, Germany

Preface

This last decade has witnessed the intensification of major conflicts in several different regions of the world. Among these, migrations from Africa and political instabilities from the Middle East generated a series of tensions in the Mediterranean area. From a geopolitical point of view, southern Europe is one of the most useful pieces of the global puzzle to maintain stability and security within and across Europe and the rest of the world. This argument is itself, enough to justify an extensive reflection about the importance of this region and its countries. Nevertheless, the scholarly focus on southern Europe has been sparse. Europe's history, tradition, political contexts, and social and economic development as well as the legacy of Greece, Italy, Spain, and Portugal, four countries with a total non-neglectable population of more than 128 million inhabitants.

The illustrative image brought by the Portuguese Nobel laureate in Literature, José Saramago, raises in his novel The Stone Raft an interesting question: What if the Iberian Peninsula would drift apart from the rest of Europe? A large shaft would divide a floating island from the continent, while the raft, shaped from the Iberian Peninsula, would drift to the unknown, en route of a collision to unpredictable places. This scenario shares an interesting starting point for sustainable development in southern Europe. Nowadays, the European Union has had large importance in planning and intensifying the decision-making processes for economic growth and development for future generations of its member states. However, just like in Saramago's stone raft, a raft is becoming increasingly formed between the north and the south. This is mainly due to the economic recession Europe has witnessed, and while central European countries such as Germany and northern Europe seem to be catching onto firm soil, southern Europe seems to be drifting in the uncertainty of its future.

The large impact of the economic recession on sustainable development is also felt strongly at the spatial level: while infrastructures built during the 1990s are still existent in many places in southern Europe, many have become over the last decade doomed to inevitable debris for future generations. Within the economic recession, the available policy support and investment in research and development have also become strongly conditioned, leading to apathy from decision-makers without financial possibilities to investment and maintenance. The European Environmental Agency has released a concerning report, stating that urban sprawl is an increasing phenomenon (EEA 2006a), resulting from a lack of planning and the inability of decision-makers to cope with the current crisis. With over 75% of the world population located in coastal zones, the conflict between urban expansion and future coastal risk brings almost surreal consequences to the future of Europe as we know (Vaz 2014). This will be especially felt in southern Europe, where the coastal activity is one of the most important economic sectors of activity, opening an additional fragility to its economic, social, and ecological systems.

The landscapes of southern European cities, once rich in wetland systems and of unique biodiversity and natural, have become on the verge of an imminent and unprecedented decline. Furthermore, coastal cities, which are rendered to urban sprawl, have majorly damaged natural heritage, archaeological landscapes (Vaz 2020), and quality of primary sectors, also facing a novel challenge besides the economic recession: climate change. However, little work has been done to assess the complexity of this threefold paradigm of economic decrease, urban change, and climate change, combined with the impacts of future cities, urban sprawl, and natural heritage. Climate change in southern Europe is affecting directly coastal erosion and will have unprecedented consequences on the city nexus as well as the loss of most historical heritage existent in Portugal (Vaz et al. 2012), Spain, Italy, and Greece. Meanwhile, the synergy between the coast, tourism, urban expansion (Vaz et al. 2011), and regional environmental change (Vaz et al. 2013) is a concerning issue in southern Europe with direct impact predictable on the diversified habitats for wildlife, ecosystems, and potential for economic growth and sustainable development. Within a changing landscape, the economic growth and the opportunities that are still available should carefully be assessed to examine the current state of coastal environments in southern Europe, as to understand the current risk and the potential still available for sustainable development in most southern Europe. Another major problem arising from the last decades of intensive investment in such countries is an emerging corruption at an institutional level, in particular in what concerns banking and building the sectors. The context readdresses the responsibility of bankruptcy and bad management to public responsibility while emerging fortunes are created offshore.

Our book is part of a long-term research project discussing the above issues, but not tackling from the start the multitude of issues deserving scientific attention and discussion. With this publication, the authors are trying to promote a critical analysis from their area of expertise while calling the attention for an emerging problem, also addressed from a spatial perception. For example, inhabitants in the region are facing the loss of rural land to urban contexts without progress, which brings very profound and severe consequences for the regional progress of these countries. Southern Europe: Major Trends and New Prospects refers to the changes in urban and rural southern Europe, mostly addressing an analysis of innovation in its cities, urbanization, and regions, to reinforce the need for a continuous reflection and involvement.

Relying on an integrated approach to assemble spatial data and extending innovation as well as sustainability to the debate of landscape and urban and rural interaction (Vaz et al. 2010), it becomes nowadays of great importance to measure as well as visualize and inform on current and future changes at the regional level. The combination of these different silos leads to the importance to assess the carrying capacity as well as the landscape and rural performance of southern Europe urban regions, cities, and rural hinterlands.

A combination of spatial analysis and advanced statistical and economic models allows assessing the challenges of the regional future of southern Europe. This depicts a clear dimension of sustainable development that must be considered in generic terms to be integrated into modern decision-making and abridge new trends for a more efficient and prosperous future. The richness, heritage, and potential at the regional level of southern Europe are tremendous (Vaz et al. 2014). By integrating and developing methodologies that pose integrated methods to protect our natural landscapes and spatial regions allows for advances in the scientific fields of social sciences and defines an important dimension for sustainable cities, regions, or nations. Furthermore, at spatial level comprehending and using methodologies to evaluate from an environmental assessment the risk of the landscape is not only an asset for the empirical changes in a context of climate change, but rather, the lessons learned in southern Europe are ones of understanding also future risk at global level, in an attempt to learn from the past and protect fragile ecosystems and biodiversity from unmanned urban growth and obsolete planning.

In this volume, three paradigms are initially labeled, corresponding to different book parts: Cities; Innovation and Landscape; Analytical Tools and Nature; Regions; and Urban Challenges. All of these aspects are fundamental in comprehending the complexity of sustainable development in southern Europe. While the first section deals with the often unmanned directions taken by a lack of strategies for innovation of governmental cities dealing with urban and rural planning, the second section expands on the analytical tools to understand the production capabilities of local governance and the impact they can have on cities and their performance. Finally, the third section puts this in perspective and approaches an integrative holistic view of the consequences of urban sprawl and the diversity of preservation of the environment as well as a landscape within the framework of regions (Vaz and Nijkamp 2015). These dimensions are intrinsically conceptual; the core objective is the assessment of the spatial analysis of the impacts of these dimensions, regarding three research clusters which form what might be considered a choice for future sustainable development in southern Europe and new paths to reconcile economic growth, nature, and landscapes within regions (Vaz 2016). The integration of spatial information and quantitative methods for regional analysis within a decision-making context is of utmost importance for southern Europe, where in recent decades and in particular current policies Europe have proposed significantly changes from a regional and a national perspective, regarding socio-economic drivers and existing policies. Southern Europe is witnessing processes of profound sociological and economic changes in which regional sciences may be of paramount interest in integrating visions of a more sustainable urban, city, and ecological future.

This book stays an open project. Our goal of altering the political power and efficiently addressing the regional governance structures is, in conscience, quite ambitious and will only be fulfilled if developed at a much bigger scale. Thus the role of this book is to inspire scholars for positive change. It is an effort of cities and regions that allows deriving the power of synergies that may robust Southern Europe in the future. It is through the hope of enthusiasm that this book poses a hypothetical view for the structural development of southern European countries that carry such a rich heritage since antiquity.

Public policy has a major role to play in southern Europe. The region calls for a permanent observatory, able to monitor and accompany its dynamics and concerns and also to find replies to several questions that have been raised by the authors during their research such as (1) Why the significant gapping trend from the rest of European countries despite the strong commitment from the European Union to reduce regional development asymmetries and modernize the governance systems of these countries? (2) What explains the persistent, low innovative entrepreneurial issue, so close to technological innovation even when regional universities do their best to profile among the best in Europe and the world? (3) Why the persistence in keeping up with corrupt political systems debiting the countries and jeopardizing the best opportunities to progress?

A set of eight chapters address these issues. The first chapter, entitled "Diversity and Country Performance," aims to launch in readers the idea of the state of the art of southern Europe's productive systems. Providing a formal presentation from the social-economic point of view, we analyze some productivity indicators that, over time, allow a perception of structural change, even at the level of emerging locational choices. These are of particular importance for southern Europe as countries have undergone profound changes due to the rapid integration into the European Union. Bringing to the whole region a fast exposure to open markets, followed by global markets, the urgent need to innovate both in terms of processes and products became evident, enabling a substantial impact on the need to welcome new production systems.

The second chapter, "Regional Opportunities in Southern Europe," draws attention to a phenomenon emerging from the observation of the initial discussion: the fact that much of southern Europe belongs to the rural world, whose foundations always sway in structural change environments. Herewith we refer to the pillars of rurality, which we consider being a scenario of modernity based on new governance systems. The title of the chapter relates to opportunities. Such opportunities emerge from the two scenarios of change we have provided and probably the only valid ones for the survival of Southern Europe. It is a space so rich in a historical and cultural legacy representing our occidental civilization's launching.

The third chapter represents the methodological continuity of the second one, "Landscape and Heritage in Southern Europe," demonstrating our argument from the previous section. The integration of geo-visualization and spatial modelling is proposed as a fundamental tool to monitor southern Europe's sustainability.

After an evolutionary approach of the territorial space, the authors introduce a consistent socioeconomic assessment in Chap. 4, "Analytical Tools from a Socio-economic Point of View." Illustrating from an empirical standpoint, they also offer a comparative analysis of different production systems in southern Europe.

The empirical analysis serves as a point of conciliation between the argument that the strategies for the development models of Southern Europe should never be planned without being based on a clear and pragmatic evolutionary analysis of the focused territories' productive systems.

The arguments of the fifth and sixth chapters entitled "Behavioral Patterns of Innovation in Lagging Regions of Southern Europe" and "Modelling Regional Innovation Patterns: The Case Study of Portugal" reinforce the set of arguments. In the first case, the evidence is provided that each territorial space can model specific patterns of different business behaviors. The sixth chapter explores that the regional dynamics of these spaces are different, although aggregable in clusters that facilitate intervention and collaboration at the networking and governance systems level.

After a detailed analysis at the level of the territories, the seventh chapter completes the specific study of the territorial area using novel geocomputational methods at the level of urbanization of agricultural areas, pointing to new emerging rurality in southern Europe, a very important concern in the context of rapid urban change in the region.

Chapter 8 closes the book with the ongoing importance of agricultural land and the richness as well as the diversity of rural land use and small towns in southern Europe. By exploring these vectors further, this chapter brings the importance of using geocomputational methods in the context of understanding the consequences of the dynamics of land use change and the possible impacts of agricultural land for the future of southern Europe.

Chapter 9 launches the comparative analysis of current land use patterns concerning the re-designation of agricultural land for urban use, permitting us to relate the results of urban variation per municipality with a variation of losses in RAN. In this sense, insights are offered on the importance of local dynamics beyond the regional-sphere of spatial-decision support systems and regional growth, converging policies, and governance that tend to have cyclical regional trends in southern Europe.

In its entirety, this book is motivated by the profound resilience of southern Europe, brought by a historical, cultural, and environmental heterogeneity that grants it a place to succeed in the future. The integration of landscape, innovation, and the impacts on land use and regional change create a functional dialogue of a story of resilience and hope where southern Europe has a unique potential to manage its growth, despite the present and future adversities.

Toronto, ON, Canada — Eric Vaz
Faro, Portugal — Teresa de Noronha

References

Vaz E (2014) Managing urban coastal areas through landscape metrics: An assessment of Mumbai's mangrove system. Ocean Coast Manag 98:27–37

Vaz E (2016) The future of landscapes and habitats: The regional science contribution to the understanding of geographical space. Habitat Int 51:70–78

Vaz E (2020) Archaeological Sites in Small Towns—A Sustainability Assessment of Northumberland County. Sustainability 12(5):2018

Vaz E, Nijkamp P (2015) Gravitational forces in the spatial impacts of urban sprawl: An investigation of the region of Veneto, Italy. Habitat Int 45:99–105

Vaz E, Nijkamp P, Painho M, Caetano M (2010) A multi-scenario prospection of urban change-the case of urban growth in the Algarve

Vaz E, Nainggolan D, Nijkamp P, Painho M (2011) Crossroads of tourism: a complex spatial systems analysis of tourism and urban sprawl in the Algarve. Int J Sustain Dev 14(3–4):225–241

Vaz E, Cabral P, Caetano M, Nijkamp P, Painho M (2012) Urban heritage endangerment at the interface of future cities and past heritage: A spatial vulnerability assessment. Habitat Int 36(2):287–294

Vaz E, Walczynska A, Nijkamp P (2013) Regional challenges in tourist wetland systems: an integrated approach to the Ria Formosa in the Algarve, Portugal. Reg Environ Chang 13(1):33–42

Vaz E, de Noronha Vaz T, Galindo PV, Nijkamp P (2014) Modelling innovation support systems for regional development–analysis of cluster structures in innovation in Portugal. Entrep Reg Dev 26(1–2):23–46

Contents

About the Authors

Eric Vaz is tenured Professor and Director of the Laboratory for Geocomputation, Department of Geography and Environmental Studies, Ryerson University. Until 2020, he served as the President of the Canadian Regional Science Association. He also sits in several editorial boards for a multiple set of key journals in the fields of regional science, development, and anthropogenic activity. With over 70 scientific contributions in his field, he was distinguished in 2012 with the nomination of "Rising Star" by the Regional Science Association International. In 2015, he received the award of the Dean's SRC Award for the Faculty of Arts at Ryerson University. Professor Vaz focuses on using spatial analysis methods, and complex system modelling approaches as well as Geographic Information Systems and Science to understand regional dynamics and integrate a better understanding of policy and the anthropocene.
Department of Geography and Environmental Studies, Ryerson University, Toronto, ON, Canada

Teresa de Noronha is Full Professor of Economics at the Faculty of Economics, University of Algarve, Portugal. She was for many years the President of the Research Centre for Spatial and Organizational Dynamics (CIEO) and Director of the PhD Program in Innovation and Land Use Management. In her international academic career, she has been Guest Professor at the Université Paris I, University of Gent, University of Bologna, and University of Toronto. Her recent research points to the frontiers of innovation and local development. She analyzes different behaviors of economic agents in a context of spatial dynamics, in particular in peripheral areas, to better promote local development. She also writes about the importance of knowledge management in the development process also in small and medium size towns. To date, she has edited and authored numerous books and book chapters related to regional and innovation economics, management in small businesses, and local development and has produced hundreds of scientific and technical papers related to such topics.
Faculty of Economics, University of Algarve, Faro, Portugal

Chapter 1
Diversity and Country Performance

Abstract This opening chapter offers a country-based analysis of socioeconomic drivers of innovation and territorial governance, bridging the significant trends in recent decades toward the optimal change of the organizational landscape. It is found that southern Europe, despite a tremendous potential of culture, landscape, history, and endogenous growth, is lagging. This leads to a thorough discussion of the south of Europe's production systems, and how they should locate themselves in line with sustainable development and regional innovation systems. It is found that urban sprawl has been a critical driver for the relocation of organizations in the south through cluster analysis. However, the exploration of new innovative systems must consider the strength and opportunity of regional sustainable development by reverting to traditional products as well as discuss the integration of sustainable drivers of agriculture and local development for efficient sustainability of southern Europe in the decades to come.

Keywords Sustainable development · Regional innovation · Spatial decision support systems · Geographical analysis · Cluster analysis · Agricultural development · Urban growth · Southern Europe

1.1 Productivity, Structural Change, and Locational Choices

The field of economic sciences has so far identified many different factors which are essential for the prosperity of humankind in society. Despite this, economics has often been accused of misconceptions and misguided behavior. This is fundamentally unfair considering that the primary goal of economics is to find best methods to manage scarcity, as well as human and capital resources offering benefits to the entirety of the system in the long run. Such a challenge, however, is not easy to undertake, especially due to the fact that economics, as a social science, is intrinsically ambiguous and subjective. Experimentation with humankind and its social abilities as well as ineptitudes may irreversibly damage society. Thus, as to avoid

E. Vaz, T. de Noronha, *Sustainable Development in Southern Europe*,
https://doi.org/10.1007/978-3-662-62177-6_1

hindrances and failures, economics must slowly progress and become very attentive to the interactions with all other sciences across the broad spectrum of scientific knowledge that use socioeconomic conclusions. The field of political and administrative sciences belongs to such a group of complementary fields. This book offers an integrated vision of southern Europe, a region which tends to have several socioeconomic, environmental, and urban-rural challenges. During the last decades, decision-makers have exercised multiple attempts to frame national governance systems in countries such as Greece, Portugal, Italy, and Spain often with unsuccessful results. By describing some of the most relevant features of socioeconomic activities in southern Europe, we aspire to emphasize its economic identity and long-term well-established social mechanisms. This shall offer a toolkit to modern administration status within the current challenges of southern Europe's status quo. Concerning economic activity, history has neglected organizing the economic actors so that each one fits a separate task but, together, all can produce the necessary goods for social welfare. In this process, not always are the interests of economic agents equal; conflicting visions require a robust regulative power with capacity and an administrative power to reinforce legislation throughout the hierarchical legal structure of regional governance.

Although economic activity appears to embody a social objective to produce essential goods and services, it interacts similarly to a living organism, composed of multiple functions, which provided an ecosystem of three domains – people, businesses, and the government – in which populations justify the existence of companies by consuming what is produced. Companies, regardless of what they produce, are responsible for making their strategic choices within different silos (profit, personal prestige, family stability, etc.), and most of the consumers' choices are driven by trends and prosperity within a capitalistic model. The state regulates, by acting better or worse and according to different political profiles, using democracy as a regulating mechanism across most of the developed world.

Within these complex interactions, the countries are presented with an enormous diversity regarding production structures, revenue distribution, or consumer patterns upon which many different governance systems must occur. Contrasts are present and asymmetries in development detected which are caused by various reasons, frequently of historic nature leading to path tendencies of future interactions. And thus, a regulatory and socioeconomic framework emerges, within the zeitgeist of needs and demands of the present that must inevitably shape the needs and demands of the future.

In such a setting, southern Europe becomes an interesting and diversified region. Despite its relatively small size, most of the arguments mentioned above occur in the complex interactions of a heterogeneous system of regulatory frameworks. For this region, history and natural resources have determined a path dependency that once represented knowledge, growth, and prosperity. In most of the southern European countries, today, however, progress tends towards financial dependency and consequent significant social struggling. A more in-depth analysis of this region is of utmost importance as to understand the tacit values of its assets and lessons learned, promoting the contextual awareness of its population while generating a robust

Table 1.1 Variation in PIB per capita

Countries	PIB per capita (PPS)		
Years	1995	2014	%
UE28 –	15 110.5	⊥ Pro 27 491.8	0.450
Denmark	19 555.5	34 226.0	0.429
Spain	13 748.7	Pro 25 021.0	0.451
Finland	16 149.3	30 280.5	0.467
Greece	12 857.2	Pro 19 938.2	0.355
Italy	18 376.2	26 355.7	0.303
Portugal	11 448.5	s 21 400.6	0.465
Sweden	19 053.0	33 706.5	0.435

Source: PORDATA, Eurostat, and the National Institutes of Statistics

regional identity. Most of all, understanding southern Europe brings a timeless beacon for modern governance and future decision-making. We defend that besides reinforcing budgetary goals, southern Europe should reinforce drivers related to the improvement of social confidence, respect, and trust, converging into cooperation, and, finally, offer a unique vision for its diversified cities.

Table 1.1 shows the variation in GDP per capita between 1995 and 2014 in northern and southern European countries. In 1995, the observed values were lower in the case of the southern European countries, but the variation rates are comparable to the ones achieved by northern countries despite the hefty budget constraints that countries such as Portugal and Greece had to face during their austerity periods of the last decade. In the last 30 years, most of southern Europe has been significantly financed for structural change, entrepreneurial activities, and education. Outcomes may cause a major concern if the result is not a clear progress occurring towards development and sustainability in such a region.

The most structured theoretical approaches on the issue of development have their beginning after the end of the Second World War, with the inclusion of factors such as "living conditions" in the concept of economic growth progressing to the construction of the Human Development Index (HDI), to an extent that improves the quality of this indicator. HDI eases the detection of asymmetric distributions or use of assets for development and emphasizes the heterogeneity and asymmetric trends of growth pushing any analytics into the paradigm of asymmetric development. This issue, however, does not only pertain to the discrepancies between developed and developing countries. We can also see signs of poverty and lack of welfare conditions in many rural or peripheral regions throughout the developed world as well as suburban areas of major cities, contracting with many cities of developed countries.

Table 1.2 brings a preliminary concept on how HDI has evolved in contrasting northern and southern Europe. It should be noted that Portugal, Spain, and Denmark were those countries to invest significant efforts in Human Development actions. However, both Portugal and Spain, departing from lower indices, were not able to obtain these higher levels seen in the northern European countries. During the 1990s, the results obtained by Portugal and Spain were surprisingly high, only followed by

Table 1.2 Trends in Human Development Index

	Year variations					
Countries	1990	2014	1990–2000	2000–2010	2010–2014	1990–2014
Norway	0.849	0.944	0.77	0.25	0.11	0.44
Denmark	0.799	0.923	0.76	0.53	0.41	0.61
Sweden	0.815	0.907	0.96	0.04	0.16	0.45
Finland	0.783	0.883	0.90	0.25	0.13	0.50
Spain	0.756	0.876	0.90	0.47	0.27	0.62
Italy	0.766	0.873	0.79	0.47	0.13	0.55
Greece	0.759	0.865	0.51	0.81	−0.04	0.55
Portugal	0.710	0.830	0.97	0.47	0.33	0.65

Source: UNESCO Institute for Statistics 2015, United Nations Statistics Division 2015, World Bank 2015 and IMF 2015

Sweden. However, the period 2010–2014 showed a significant break due to the global crisis and brought a lagging process for the region, both in regard to GDP and HDI.

Regretfully, these new activities of international emerging markets block the solutions of more sensitive economic activity components within countries, companies, or people. Hence, the role of political economics and public policy has become essential in the analysis of the most appropriate choices depending on the critical characteristics of the situation. They are different for each case and must be assumed to be irreversible.

Discussions related to distribution of wealth and inequality have recently been brought to public attention, after Angus Deaton, Nobel Prize laureate in 2015 (Deaton 2003), for assessing the key issues of consumption, poverty, and welfare. This arrives just when the scientific community discussed inequality (Atkinson and Bourguignon 2014; Piketty and Goldhammer 2015). His research concentrates on the consumption patterns that would not reflect the aggregate national consumption in *The Great Escape: Health, Wealth, and the Origins of Inequality* (2013). The author claims that although, in general terms and around the world, it had been possible to observe gains in health and well-being arriving from GDP growth, over the last decades, there are many people falling short of such improvements. This work has the great potential to warn to the hidden performances that macroeconomic data does neglect and fail to recognize. In our view, this rational is as valid for individuals within the society as for regions. Even though southern Europe has a series of strong development indicators, a lack of regional progress leads to the circular dynamics of peripheral governance, and insufficient market mechanisms. Triggered by such circumstances, several factors are explored as underlying reasons:

1. One of the major determinants (or consequences) of the earlier described situation is that despite the high investments in education and labor specialization of the last decades, low productivity persists in most of the southern European

Table 1.3 Labor productivity per hour in some European countries

Countries	Labor productivity per hour (UE28 = 100)	
Years	1995	2015
UE28	100.0	100.0
Denmark	131.1	126.6
Spain	107.8	Pro 99.8
Finland	108.4	105.7
Greece	74.2	Pro 68.2
Italy	123.1	100.6
Portugal	64.1	s 68.8
Sweden	119.0	114.3

Sources: Eurostat and PORDATA

countries – see Greece and Portugal in Table 1.3. This table further illustrates the stark contrast between these countries and Spain or Italy, countries that are clearly moving out from a consolidated industrial economic structure towards a new technology and a service-based economy. The global crisis of the last decade did not permit any country (listed in the table) to obtain higher productivity rates, with the exception of Portugal, showing a slight increase but still lagging regarding others.

2. Factors related to structural change and innovation direct the discussion into a dynamic and systematic approach opening the potential of societies to interfere in the processes and transforming their intrinsic path dependencies.

The empirical findings have shown that, throughout history, technological change has always been the great engine of economic dynamics and prosperity of firms in countries. Numerous scholars investigated the causes of growth, trying to formalize and model these relations.

For instance, investment has long been considered a major factor in economic growth. Recently, the importance of technological change has been emphasized, and much of the related theoretical approaches suggest that growth depends upon sets of tangible assets through determinants of innovation as referred by Solow (1956). Since then, much has changed in his proposition. Arrow's model (1962) introduced the concept of "learning by doing" as a determinant of technological development. Later, Lucas (1988) introduced human capital as a main factor of technical change, and in the late 1980s (1986), Romer considered technical change endogenously determined by research. Spillover effects from such determinants have been explored in the Marshall-Arrow-Romer model and presented by Acs and Audretsch (1984), Acs (2002), and Audretsch (2002). In summary, nowadays, technological innovation is perceived as the result of an extensive knowledge base, generated by multiple inputs, integrating intangible assets as well as the preceding premises of growth.

Within the process of growth and generation of wealth, one of the vectors of technical change, innovation, and knowledge converge is the organization's capacity of learning. Although intangible, such major driver can be related to internal or external factors to the firm, particularly if human capital can serve to perceive the

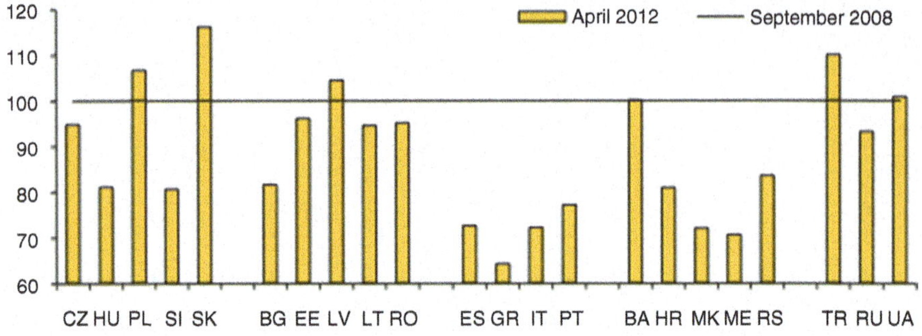

Fig. 1.1 Development of industrial production. (Source: Landesmann 2015)

nature of such learning capacity. Social learning capacity is hardly quantifiable, but proxies may be used to estimate, for instance, the organizational capacity to learn. Examples have been proposed by Noronha Vaz and Cesário (2004, 2008). Their research used features such as multiple characteristics of the top managers, skills, and training for the workforce or other aspects such as interactions with suppliers, customers, industry associations, and public support bodies to determine the firms' capacity to innovate, measuring their true contribution to national growth and welfare.

As growth models depict technological change as well as the multiple factors that generate growth as determinants of the production function, a clear formalization of economic growth is still necessary. From a non-dynamic perspective, the economic activity is very well portrayed in a structure of relationships between agents in which the output is the final value of all productive activities. At the outset, the system can be simplified and disaggregated into sectors (agriculture, industry, and services) which in turn are sectioned sub-sectors. These are public and private endeavors that contribute to the final products, each with the task of producing one or more products (goods or services). It is a structure whose complexity results from the fact that each company relates to the acquisition of other intermediate goods and services leading to continuous flows of transactions (including financial). This increases and absorbs constantly the consumption as well as the household savings (Noronha Vaz 2011)

In the case of southern European countries, it is worth observing their economic structure to find out some reasons for their lagging behind concerning growth and better perceive the potentialities for structural change and growth. The past has determined a very low propensity for industrialization in the four countries considered.

Figure 1.1 concerns three different periods (2001–2004, 2005–2008, and 2008–2011) and compares the development of industrial production in countries of southern and eastern Europe. For all these four countries, the manufacturing sector performed poorly concerning its contribution to overall GDP during the pre-crisis period. In such cases, GDP growth was almost entirely based on the service sectors or construction as can be observed further in Fig. 1.2, for which manufacturing (C),

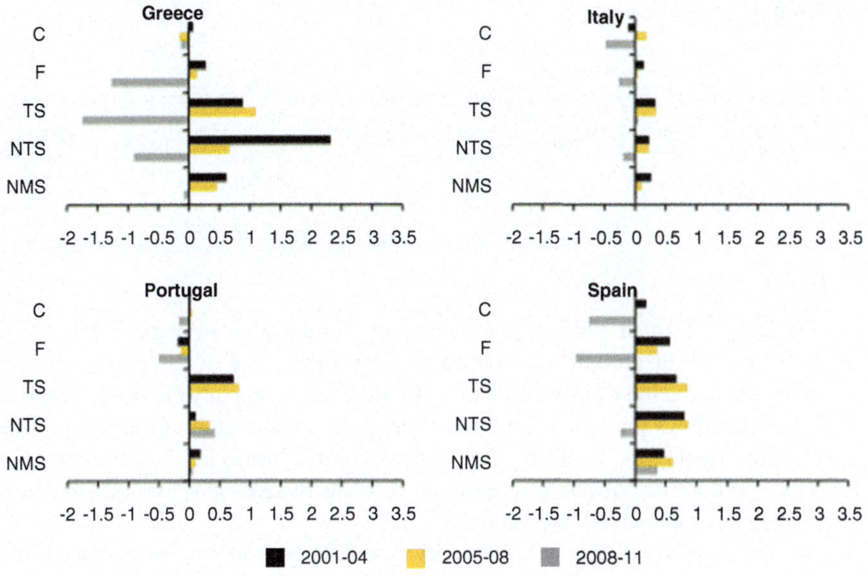

Fig. 1.2 Development of industrial production. (Source: Landesmann 2015)

as the classic tradable goods sector, construction (F), and wholesale and retail (G) sector and other market service activities (H-N) are shown.

In the search for the possibility of convergence within Europe, Landesmann (2015) compares southern to eastern European economies. His work develops an extensive data analyses, and his conclusion points to very distinct differences between GIPS and eastern European economies. For example, Landesmann (2015) confirms that structural distortions had summed up shortly before 2011, with more severe consequences over the decrease of manufacturing for southern European economies of Greece, Italy, Portugal, and Spain, while the growth pattern was more balanced in the northern European countries. Nonetheless Baltic countries and Slovenia also experienced a strong swing away from manufacturing just formerly to the outbreak of the crisis.

Further, the author refers that, in the case of a postcrisis adjustment, manufacturing was also strongly negatively correlated in many countries, challenging any process of readjustment supported by the trading sector, crucial to solving current account debts and the high external debt. Landesmann's research also confirms significant differences between southern and eastern European economies such as a more pronounced reduction in manufacturing before the crisis in the southern than in eastern economies.

Countries' economic structures hang on to their production systems which, to avoid collapse, must be able to reproduce in successive cycles of production without tending to a situation of steady state and, eventually, stagnation of savings or consumption. One can assign each company a life cycle resulting from the combination of the different life cycles of its products which, in turn, feed the continuity of the welfare of regions or countries.

Table 1.4 Central government debts

Countries	Years 1980	1990	2000	2010
Sweden	38.2	39.6	56.9	33.8
Spain	14.3	36.5	49.9	51.7
Italy	52.7	92.8	103.6	109.0
Greece	–	97.6	108.9	147.8
Portugal	29.2	51.7	52.1	88.0

Source: OECD Statistics and World Economic Outlook data 2013

The life cycle of a product (or company), similarly to biological life cycle, presents a phase of growth, maturation, and the final stage of disappearance. By worrying about the life cycle of each of its products, the company chooses a survival strategy that allows the extension beyond the point of maximum production. This effort of prolonging the maturity stage of its life cycle brings a focused strategy of renewing productive skills and targeting investments towards new products and new processes and is known as innovation.

This focus on innovation given by society and its aspiration of prosperity could take us to confirm the earlier argument on diversity: technological change demands diversity in the productive measures, and, hence, the economic life diverges at the multiple levels of our social existence. However, few have been the ways found to integrate the multiple productive modes in a sole concept of wealth, and even fewer have been the accepted perceptions of prosperity and well-being. The investigation about what is taking place regarding economic development and sustainability in southern Europe should serve to expand tolerance towards what prosperity and well-being in society is and how different productive modes are also called to improve the existing modus operandi.

Dramatically amplifying the lack of structural adjustment of Portugal, Greece, Spain, and Italy to fast growth, the last financial crisis represented for most of these countries a severe downturn towards cohesion within European countries and accentuated, even more, a historical path-dependent position for the region of southern Europe. Greece, Portugal, and Spain, with those structural debilities earlier described, were unable to financially overcome the negative effects of the post-2008 financial international collapse. As the European Union and the International Monetary Fund encouraged southern European nations to follow austerity, in exchange for financial assistance, slowly but surely the economic structural change stagnated further, accruing inequality, rising poverty, failing institutions, and increasing the lack of trust in the political class.

Zamora-Kapoor and Coller (2014) wrote an excellent paper concerning governmental debts as principal causes for immigration, as well as failures in institutional difficulties in responding to the ongoing crisis. Table 1.4 shows the evolution of historical central government debts to which strong austerity measures have been used. In 2011, the consequences of such measures were significant average household income reductions in Greece (by 14%), in Portugal (by 7%), in Spain (by 5%), and in Italy (by 3%). Moreover, policies produced directly or indirectly new forms of inequality.

These conditions of international conjuncture affected profoundly the progress of the southern European countries with two major unexpected and irreversible structural impacts: Firstly, after having incurred in costly advanced training, the countries face now a significant "brain drain" immigration as richer countries, such as Germany or the United Kingdom, can better profit from specialized skills at zero initial cost. Secondly, a profound mistrust in the political class is growing because of both the states' inability to solve urgent problems and lack of transparency and ethics occurring within the political classes of those countries (Zamora-Kapoor and Coller 2014).

Finally, the third set of factors is particularly important and falls into the scope of locational choices. Acting in their different roles, economic agents such as businesses, which are producing marketable goods and services as the main source of income, or families, considered consumer or saving units, have locational preferences related to their specific functionalities; their mobility capacity leads them to the formation of larger or smaller settlements. There are multiple factors that can lead to different choices for business location: the way productive sector companies operate, their size, and their stage of maturity are just a few. In both cases, transportation costs are always constraints of location dynamics for any company as well as accessibility or the proximity to markets for raw materials and consumers.

The productive sector can motivate the company to locate near the sources of energy production, for example, or the supply of raw materials; the large-sized company with the intensive use of human resources, on the other hand, may be forced to locate where cheaper labor is present. Also, the degree of maturity of the companies related to its life cycle can fix different strategies, thereby creating changes in levels of their capacity for sustainability.

In the short term, the objective of the entrepreneur is to minimize costs and maximize revenues for the company. However, in the medium term, the entrepreneur needs to dominate the price mechanisms and is prompt to change its initial location. If faced with decreasing production, the company may need to plan to increase its market share and look for alternative locations which provide it with a wider market share. Only then does it reallocate in search of more favorable conditions to its objectives and strategies.

Frequently, the economic agents also find agglomeration advantages out of their location in specific places within the geographical space, encouraging specific behaviors from organizations. Location or agglomeration economies provide better environments for the economic development. The different agglomeration types design different social spaces in which public and private corporations coexist and relate within certain territories. These should also be considered instruments of development as they directly affect the welfare of populations.

The fact that locational advantages are common to many economic agents leads to an increasing spatial agglomeration and brings specific urban patterns where human intervention has contributed to different sociocultural path tendencies. A stark example is found in the contrasting realities of the rural and urban world, where opposition of the key characteristics exists, but a complementary process for development is harbored.

The locational behavior of consumers is based essentially on two types of factors: the proximity to employment and housing supply and demand that is determined by the disposable income.

However, the spatial distribution of economic activity may influence other determinants that steer consumers to different choices of demographic concentration. This applies to the use of common goods and services (public goods, for instance), usually essential to the well-being of the people and provided by the state to increase the effectiveness of management, in particular central places.

Given urban prosperity, cities (which are the centers for the privileged location of economic activities in space) change their morphology and restructure spatially. The transformation of the productive sectors and new economic agents' locational choices result in new settings for the rural, industrial, and urban spaces. Crowding regions also raise issues concerning the different questions related to human intervention of geographical space where balance among natural, social, and economic systems should be the ultimate goal of sustainable development. This paradox is associated with the sustainability of the production process and has raised different viewpoints over the past decades (Vaz et al. 2013). The arguments that support economic growth and that are based on the ever-changing technology to support it are juxtaposed by pressing positions of social restrictions that demonstrate the impossibility to consume new products, given the difficulty of instilling in the consumer continuing new needs and, in the long term, scarcity of natural resources. Fragile ecosystems, such as those prevailing in the Mediterranean region, should take special care of this concerning paradigm.

1.2 Innovation and Cities: From Old to New

So far, new processes, new organizations, and new market decisions are necessary to innovate, but, at the same time, all these may delay the decrease of production preventing the product's life cycle from entering the final phase of its existence. Innovation is, thus, considered to be the optimal choice for the survival of organizations. The more the economy becomes competitive, the greater the need to include innovation in the production processes. The phenomenon results from the application of new ideas to production processes and results in technological and organizational changes. In the long run, this becomes a mechanism of cause and effect of growth and social change.

In recent decades, the economics of innovation became a branch of economics and management to study and better understand the effects of technological change (Vaz et al. 2014a; Domínguez et al. 2014). Knowledge has been identified by most scholars (Vaz et al. 2015) as being one of the major determinants of innovation.

As a concept, abstract knowledge requires the application of further efforts in training human resources. However, knowledge is also a fact that cannot be reduced to the simple relationship of teaching and learning. Nowadays, companies carry a load of historical and cultural traditions or country-dependent legal frameworks, encouraging certain forms of knowledge transfer, almost intuitively promoting the observation of specific experiences, and learning with these. This is the basis for tacit knowledge, a form of knowledge closely linked to geographical spaces, territories, and its unique identities (Noronha Vaz and Nijkamp 2009b; Vaz and Nijkamp 2015).

Not only organizations but also populations may benefit from tacit knowledge, according to their sociocultural and institutional contexts. Even though it may be difficult to forester tacit knowledge in the short term, literature demonstrates that tacit knowledge and regional sustainable development are highly correlated (Noronha Vaz and Nijkamp 2009a).

Codified knowledge is another form of knowledge. It promotes specific and directs new production processes by representing a specific sector's know-how and requires in-depth expertise in certain areas of knowledge. This kind of knowledge perpetuates itself with a multiplier effect that results from adapting to new operational methods and techniques where dissemination and coding of information are a key part (Pinto et al. 2015).

These two dimensions of knowledge complement each other, while it is largely accepted that the abundance of tacit knowledge facilitates the seizure of codified knowledge and its reproduction.

Technological change and innovation are the most important themes in the literature focused on growth. The systemic approaches and models argue that the learning modalities for promoting knowledge are the best contributions to efficient results regarding productive innovation. Learning, as a collective and interactive process, permits expansion of knowledge, whether it is tacit or codified knowledge.

From this point of view, the dynamics of growth and development become a context in which knowledge production and management are paramount. From an economic perspective, knowledge is an asset to be produced and used; its management obeys to rules of supply and demand. The provision of knowledge or its production is scattered by sources in many different scientific fields. Their creation processes and dissemination flows are concentrated in very specific locations, benefiting from many specific economic advantages resulting from preferential location and fast dimension as any other produced asset.

The democratization of societies and the emergence of new information and communication technologies (ICTs) expanded broadly in recent decades due to easier and more available means of knowledge transfer. This is the reason why knowledge management plays a major role in directing knowledge flows throughout the channels that warrant important assets to be ubiquitously available.

As to observe how knowledge is progressing in southern European countries, we have chosen a few indicators and compared these with other European countries as follows: Internet access per 1000 inhabitants, the evolution of PhD per 100.000 inhabitants, expenses in R&D in % of GDP by sector, and, finally, companies' Internet connection as percentage of total.

Table 1.5, representing the Internet access per 1000 inhabitants, shows that despite the huge broadcasting effort among all the population, Italy, Spain, Greece,

Table 1.5 Internet access per 1000 inhabitants

Countries	2014
EU28	–
Denmark	420.4
Spain	276.3
Finland	321.4
Greece	290.7
Italy	236.4
Portugal	(275.0)
Sweden	331.0

Sources: UN, Eurostat, DG CONNECT, National Institutes of Statistics, PORDATA

Table 1.6 Evolution of PhD per 100, 000 inhabitants

Countries	PhD per 100, 000 inhabitants	
Years	2004	⊥ 2014
EU28	23.7	⊥ Pro 35.0
Denmark	20.8	53.8
Spain	26.1	31.2
Finland	38.2	51.4
Greece	16.5	19.6
Italy	19.5	23.0
Portugal	12.0	(51.5)
Sweden	43.5	52.5
Norway	24.2	40.8

Sources: UN, Eurostat, DG CONNECT, National Institutes of Statistics, PORDATA

and Portugal remain far behind northern European countries. This is not only related with the income distribution but is also a result of different cultural boundaries. As we can see in Table 1.6, related to the evolution of PhD per 100.000 inhabitants, the efforts done in southern European countries to promote advanced research are significant. One should note Portugal's remarkable progress. When observing Table 1.7, expenses in R&D in percentage of the countries' GDP by sector, a first conclusion is that such expenses have increased significantly in the public sector but have been directed to higher education, followed by entrepreneurial investment, mostly in northern countries. Portugal shows data representing a faster growth when compared to other southern European countries, despite the serious economic crisis post-2011. Our last growth indicator, reported in Table 1.8, shows companies' Internet connection as percentage of total allows us to conclude about those sectors per country with higher capacity for modernization. Indeed, this is an excellent indicator as it shows the effort of the countries per sector to enter in the international networking, being clear that all the countries can provide a significant effort to do so. Portugal is again a surprising case of departing from a very unfavorable position to a situation of significant advancement in the sectors of commerce, tourism, as well as transportation.

Table 1.7 Expenses in R&D in % of GDP by sector

Countries	Total		Enterprises		Public sector		Superior education		Private nonprofit organizations	
Years	1995	2014	1995	2014	1995	2014	1995	2014	1995	2014
EU28	x	(2.3)	x	(1.29)	x	(0.25)	x	(0.47)	x	(0.02)
Denmark	1.79	(3.05)	1.03	(1.95)	0.30	(0.07)	0.44	(1.01)	0.02	(0.01)
Spain	0.77	(1.22)	0.37	(0.64)	0.14	(0.23)	0.25	(0.35)	0.01	(0.00)
Finland	2.20	3.17	1.39	(2.15)	0.37	(0.27)	0.43	(0.73)	0.01	(0.02)
Greece	(0.42)	(0.83)	0.12	(0.28)	0.11	(0.22)	0.18	(0.32)	0.00	(0.01)
Italy	0.94	1.29	0.50	(0.78)	0.20	(0.19)	0.24	(0.35)	x	(0.04)
Portugal	0.52	1.20	(0.11)	(0.59)	0.14	(0.08)	0.19	(0.58)	0.08	(0.03)
Sweden	(3.13)	(3.16)	(2.34)	(2.12)	0.12	(0.12)	0.68	(0.92)	0.00	(0.01)

Sources: UN, Eurostat, DG CONNECT, National Institutes of Statistics, PORDATA

As observed in the past analytics, Europe comprises a heterogeneous profile which tends to accentuate the distance among European countries and regions. Southern regions had to deal with high unemployment and shrinking populations, independently from growing efforts invested towards prosperity and sustainability, a condition which seems not to be unrelated to the process of knowledge creation and diffusion. A closed observation into the historical path of these countries illustrates how education has been differently emphasized during the past centuries across Europe and how this characteristic represents a major track of regular path dependency. A second consideration is globalization which affected the entire eurozone (Noronha Vaz 2004) and the increased competition from Asia that, still today, demands for a full reshaping of the political, economic, and social structures of most of these countries. One of the major elements able to sustain regional instabilities resulting from the global trend is urbanization. Multiple considerations are underlining the role of cities, and their autonomous urban governance, as possible economic drivers for change and innovation, providing citizens' awareness is taking place (Noronha Vaz et al. 2013).

Thus, across Europe, many towns have deindustrialized (Noronha and Vaz 2015) and shrunk or suffered from escalating unemployment and low productivity calling for new industrial activities. Southern European regions have been affected by this situation and remain hopeful that regional or local actions can promote their scenic beauty and landscape or cultural assets, promoting tourism as one of the best alternatives to proceed towards structural change. The impacts of the choices remain to be seen in the coming decades. Whether this is a clever choice is still to be seen. To become smart cities is a different strategy that towns are following to develop further as well, although conditions for progress, in general, frequently result from external determinants impossible out of the control of such cities, regions, or countries, wherein particularly smaller cities are highly capable to be resilient and adopt smart technologies quickly and within their cityscape.

Table 1.8 Companies' Internet connection as percentage of total

Countries	Total		Manufacturing		Building		Commerce		Tourism		Transportation		Housing and other services	
Years	2003	2015	2003	2015	2003	2015	2003	2015	2003	2015	2003	2015	2003	2015
EU28	x	97	x	97	x	97	x	97	x	99	x	97	x	97
Denmark	97	100	§	100	93	100	§	100	§	99	§	99	§	100
Spain	82	98	81	97	74	99	88	99	89	100	84	99	83	98
Finland	97	100	98	100	94	100	97	100	97	100	91	100	100	100
Greece	88	87	82	93	84	71	89	92	83	100	90	95	97	82
Italy	83	98	82	99	83	99	82	99	90	99	85	97	86	94
Portugal	70	98	64	98	65	98	73	100	88	100	76	100	85	100
Sweden	95	98	97	98	94	99	96	99	100	99	87	95	96	99
Norway	88	98	90	99	92	100	83	98	84	100	88	98	96	95

Sources: UN, Eurostat, DG CONNECT, National Institutes of Statistics, PORDATA

1.3 Production Systems in Southern Europe

Currently, most capitalistic economies must understand that two production models underpinning today's economic activity coexist. The first model has emerged directly from Fordism and bags conditions of strong industrialization of production, whenever the intensification of corporate capital and mechanization of processes take place. In such a model, the productive activities are characterized by reducing prices of large-scale production. Products with high standardized levels and much specialization generate more benefits if produced in locations where labor is available, adding to increasing the productive capacity of businesses and the achieved economies of scale. As described by Noronha Vaz (2011), this model avoids transactions and reduces the cost of vertically integrated production processes.

Given the tendency for large-scale production, in which case the relationship is governed by an increase in the spatial division of labor, three situations may occur: (1) The regions are important and constructive environments for technology and research; (2) the regions attract well-qualified skills and display production activities increasingly more profitable; (3) or the regions have unqualified skills and, thus, receive less profitable activities tending to decline. At this point, it should be emphasized that, from a microeconomic perspective, the competitive advantage of companies starts by the reduction of production and labor costs, compelling companies to choose strategies that may impose relocation within geographical space.

The second model, defined by segmentation and flexible specialization, is a model that is wide-spreading in the productive systems of the more developed world due to the increasing labor costs and the rising use of information and communication technologies. Its application shows that the organization of production stands for a very close link to the market and results from a systematic breakdown of the production processes. For the system to work towards a set of continuous processes, the requirements relate to cautious standardization, just-in-time practice, and the creation of functional business networks, interdependent product design, and manufacturing.

This last model, also defined as post-Fordist, presents business management systems and industrial production as very complex, both from a technological and organization standpoint. In this sense, the relocation of businesses or its mobility is difficult, removing the sudden increase of regional unemployment. The sharing of knowledge and the maintenance of highly qualified personnel, as well as companies' flexibility to acquire more knowledge and develop new production processes, become the key for success.

Currently, the developed world is undergoing a transformation context in which it is possible to witness the pressure of both these models. Several fluctuations in employment and entrepreneurship result from their symbiosis, and countries differ in their integration of a clear result in cost reduction and advantages of growth. As such, innovation is no longer an individualized process of empirical application and discovery as it is becoming a collective learning process that goes further to boundaries of technological research and development. In the integration of a production model to the other, new emerging social forms, the process of business innovation appears that, coupled with technological globalization, requires collaboration strategies with companies and research centers throughout its networks and evolving partnerships.

This new vision stresses the characteristics of organizations as determinants for more efficient environments and the interactions between them and a multiplicity of local actors (businesses similar or related, suppliers, clients, institutions, R&D, and training).

The theory of entrepreneurship favors the innovation processes dedicated to specific niches, particularly when coupled with flexible and informal organizations.

Noronha Vaz and Cesário (2008) relate the quality of entrepreneurial management not only with the characteristics and structure of companies but also with the external conditions of the surrounding such as know-how, level of qualification, acceptance of institutional structures, dynamics of the economic activity, and level of industrialization.

A system of reciprocal influence helps to increase flexibility and reduce the optimal size of production units, electing SMEs (small and medium enterprises) as increasingly competitive organizations. These tend to occur more frequently in central areas where conditions of services and resources offer better opportunities for innovation and differentiation.

When such environmental conditions are adverse, especially in rural or peripheral regions, SMEs tend to specialize and search for synergies. By doing so, they can benefit the local environment, including other companies within the system for cooperation and learning.

Despite these generic tendencies, enterprises often look for a set of discrete locations for their manufacturing plants, either in small cities or in the suburban regions of the central places, to avoid direct competition between organizations concerning labor, a factor that could make scarce highly skilled labor costlier. As more firms focus on core business and become involved in more complex extra-business networks, competitive efficiency becomes more important to individual enterprises as they begin to rely on other firms to absorb specific stages of the production process. On the other hand, this transition is changing the cost structure of enterprises (Eliasson 1990), where production and assembly costs are being reduced to make room for technological advances. While the internationalization of products causes a rise in costs related to external marketing and logistics, better transportation conditions and the adoption of the just-in-time production system, on the other hand, generate a decrease in transportation and stock management costs. Greater commitment to the understanding, value, and importance of core business has increased costs related to research and development. To assess southern European regional capacity of attracting entrepreneurial activity, 43 regions from Portugal, Greece, Spain, and France have been chosen and variables clustered to match the main factors of regional dynamics. Some of those have been selected due to their positive contribution to regional attraction of firms (Noronha Vaz et al. 2003)

Variables such as the quantity and quality of route accessibility, workforce, workforce flexibility, number of staff, R&D expenditure and jobs, the distribution of gross added value, and gross domestic product per capita were considered as important determinants to characterize the different regions. The variables were assessed in relative terms to eliminate the effect of different regional dimensions regarding geographical, demographic, and economic conditions. The railway network density and the motorway network density have been used to determine the number of access routes available. Road density was considered a suitable variable as well. To better understand the type of workforce involved, the ratio of active population over total population was examined together with population density and non-active population ratios regarding quality labor; several variables have been created such as the number of students attending school at each stage of the education process in relation to the total population. The contribution provided through R&D considers the expenses as the percentage of GDP in regards to the affected region as well as the percentage of employment. Further, a distinction was made between the origin of the job and R&D expenses (business sector, state, and higher state education).

Finally, we use the gross added value distribution as an indicator of the degree of current regional development, that is, the situation various regions find themselves in, for further development.

By applying a cluster and discriminant analysis, it was possible to determine which variables contributed mostly to the group differentiation.[1] Table 1.9 shows that the first cluster comprises Greek regions except for Attiki, the Spanish Balearic region, and the Portuguese regions except Lisbon and Vale do Tejo. The second cluster consists of the regions Attiki, Lisbon, and Vale do Tejo and the Spanish regions except for the Balearic Islands. The last cluster represents the French regions.

The mean values for each of the three clusters show that cluster 1 corresponds to lagging regions that lack the capacity to develop quickly and sustainably. Cluster 2 represents those regions that are more developed and have greater potential for development than those in cluster 1. Finally, cluster 3 joins those regions that have already reached a high level of development and will easily attract dynamic firms.

The state and institutions of higher education belonging to these regions replace those firms that do not yet have a robust capacity to invest in this area, such as developing technological centers. R&D employment is in line with some expenses carried out, though the higher education sector reveals excessively high, possibly because of lack of its efficiency. Regarding education, these regions display good results concerning the number of students that attend higher education. The values for elementary and intermediate stages of education are satisfactory, an apparent

[1]To compare the proximity of all variable values, we used the squared Euclidean distance. According to this technique, the distance between two cases (i and j) is defined as the sum of the squares of the values of i and j for all variables.

The aggregation method used was the hierarchical agglomeration. Based on clusters, a discriminant analysis was performed.

Table 1.9 Clustering some southern European regions

Cluster 1	Cluster 2	Cluster 3
Anatoliki Makedonia, Thraki	Attiki	Champagne-Ardenne
Kentriki Makedonia	Galicia	Picardie
Dytiki Makedonia	Asturias	Haute-Normandie
Thessalia	Cantabria	Centre
Ipeiros	Pais Vasco	Basse-Normandie
Ionia Nisia	Navarra	Bourgogne
Dytiki Ellada	Rioja	
Sterea Ellada	Aragón	
Peloponnisos	Madrid	
Voreio Aigaio	Castilla-León	
Notio Aigaio	Castilla-La Mancha	
Kriti	Extremadura	
Baleares	Cataluña	
Norte	Comunidad Valenciana	
Centro (P)	Andalucia	
Alentejo	Murcia	
Algarve	Canarias	
Açores	Lisboa e Vale do Tejo	
Madeira		

Source: Noronha Vaz et al. (2003)

pointer that the failure rate is not high. Workforce flexibility is somewhat average, supporting the idea that the level of flexibility remains weak and needs to reach the same levels of more developed regions. Concerning per capita GDP, these regions have shown to be the intermediary, illustrating an average level of development. Finally, the gross added value distribution shows a very weak proportion of the primary sector, but a strong weight of the tertiary sector (almost twice as much as the secondary sector), evidence that GAV is nearing values from more developed regions.

Cluster 3 is comprised of those regions that offer quality road network accessibility, railways, which exhibit less costly in the transport of goods and generate fewer traffic issues. France possessed a higher number of motorways and has shown to maintain railways as a traditional and efficient form of transportation and network. Intriguingly, the population density is low. Still, if we consider the fact that these regions surround Île-de-France, which includes Paris, this low population density results from the capital's gravitational pull. Concerning R&D expenses, these are significantly high, mainly arising from the private sector. It is also due to this that most entrepreneurial companies choose their location, focusing on competitive product quality and innovation. There are advantages when private sector develops R&D instead of the public sector, since these are bound to be performed more efficiently and rationally in research areas of greater demand. In turn, these regions are characterized according to a better-qualified labor force and thus naturally circumvent the need of hiring an excess of highly qualified workers as required in cluster 2 regions. As such, with exception of lower high school education, these regions present average figures for the different education stages and can be tied to the low failure rate in the educational system. These regions also seem to prefer more technical education, whereby more job specializations are available and job responsibilities delegated earlier to young people without the need to attend higher education. Supporting strong qualifications and job specialization in these regions is the high flexibility that is evident. Concerning per capita gross domestic product, this is naturally superior, while the distribution of gross added value is identical to those regions belonging to cluster 2. The small difference weighing the primary sector is related to the sparkling wines weighing heavily in the Champagne-Ardenne region.

The results confirm those obtained by Quévit (1995), who questioned the classical economic approach of European development completely biased by the creation of advantages of global competitiveness and access to international markets, over 20 years ago. Both studies indicate that technological change, nontechnical economies of scale, and industrial cooperation will have a troublesome effect on the productive context of less favored European regions. Suggestions for regional development indicated that traditional industrial regions should promote strategies based upon intra-industrial activities and the lagging regions. Strategies oriented to inter-industrial contexts.

1.4 Conclusions

Urbanization has been the best way society could find to protect itself from external aggressions. This has been achieved by skills to increase safety and welfare to humankind. While urban growth is the current town, some urban regions shrink and disappear over time. Those surviving develop further into strategic locations for the growth of wealthy regions or nations and contribute to generate increasing spatial advantages while integrating specialized and diversified production systems.

A great part of the urban sprawl has been due to its hinterlands which always played a major role in support to development. These regions provided food and other assets, such as labor in Europe, even much before the Industrial Revolution. Hinterlands and its rural roots contributed to the basic foundations of the present capitalistic systems across the world and in Europe, where its dependent social structures still last in its southern regions. Also, here, history clearly defined the role of agriculture in the past development of regions, either by creating a frequent productive model of self-consumption or, more recently and in some cases, by introducing the industrial activity in the rural world and exploring the agribusiness generating different functionalities economically independent from their nearby located towns. This is specially the case of southern Europe, where the richness of the historical landscape has been a functional source of economic prosperity and historical heritage tourism (Vaz et al. 2012; Vaz 2016).

For an extended period, it seemed that cities and hinterlands would tend to grow apart from each other but together with the sturdy world of industrial Fordism. Its assets are produced in scale and sent away, as far as trade would allow. Thus, some hinterlands are structured on multinational corporation bases and became areas of high productive agriculture with not much population along extensive areas. Others, on the contrary, populated by non-skilled agricultural labor force are kept behind in the vicinity of urban areas and diversified its activities into traditional manufacture and related industries as lastly noticed in southern Europe.

In general terms, the generalized globalization trends seem to merge urban areas and their peripheries into a complexity of patterns of multitasking and multifunctional socioeconomic systems. This new phenomenon calls for a more comprehensive approach related to the peripheries across the world and a new understanding of the diverging concepts of the rural and the urban worlds. Southern Europe offers many examples, and its intricate systems may supply lessons from cultural transitional areas, where two or more cultures converge and interact, revealing cultural and linguistic aspects from each other. Acceptance of diversity may represent an increase in cultural capital and higher levels of social resilience, due to the different aspects of accumulated knowledge assets, especially during periods of systemic stress.

This book intends to enhance the debate of the major trends and new prospects for southern Europe, emphasizing its natural diversity and heritage as a contribution to social resilience and sustainability. The arguments describe a structured form and aspire to understand the crucial sets of human social behavior and progress, based upon few key concepts such as segmentation and specialization of the productive process, or the new instruments of knowledge management, possible new avenues that prospect the sustainability and growth of southern Europe (Vaz et al. 2014b).

The theoretical framework is not consensual. Some scholars and decision-makers doubt that a recovery of lagging peripheral regions may be possible, unless at increasing governmental costs and strict control of policies. Others suggest that outstanding intervention can reverse processes and transform the participation of such locations into true partners of the global chains and social change. In this book, southern European legacy is enhanced by employing concepts of productive systems, efficiency, and innovation. This is explored in such a way that their awareness and values should promote innovative solutions which matter to explore for a balanced and equitable development across the Europe and the world. Having this in mind, the undertaken analysis illustrates identities, characteristics, skills, and potentialities of southern Europe.

This book provides both a scientific survey on the underlying theoretical framework and a positive view of southern Europe as it is a historical construction of mostly resilient peripheries in a struggle to use skill and experiences to surmount difficulties. The authors intend to be able to claim for a new and more sustainable region. The theoretical and scientific focus of this book insists on the need for new innovative processes and products as tools for development. Furthermore, it emphasizes that by promoting networking, locals can recombine natural assets and innovation more efficiently and sustainably.

References

Acs Z (2002) Innovation and the Growth of Cities. Edward Elgar, Cheltenham

Acs Z, Audretsch DB (1984) Small Business in Industrial Economics: the new learning. Revue d'Economie Industrielle 67(7):21–39

Arrow KJ (1962) The Economic Implication of Learning by Doing. Rev Econ Stud 29(3):155–173

Atkinson AB, Bourguignon F (2014) Handbook of income distribution. Elsevier, Social Science

Audretsch D (2002) The Innovative Advantages of US Cities. Eur Plann Stud 10(2):165–176

Deaton A (2003) Inequality, and economic development. J Econ Lit 41(1):113–158(46)

Domínguez JA, Noronha TD, Vaz E (2015) Sustainability in the trans–border regions? The case of Andalusia-Algarve. International Journal of Global Environmental Issues 14(1–2):151–163

Eurostat, DG CONNECT

Landesmann MA (2015) The new north-south divide in Europe: can the European convergence model be resuscitated? In: Fagerberg J, Laestadius S, Martin BR (eds) The triple challenge for Europe economic development, climate change, and governance. Oxford University Press, Oxford

Lucas RE (1988) On the mechanics of economic development. J Monet Econ 22-1988:3–42

Noronha Vaz T (2004) The environmental context for small firms in the EU. In: Noronha Vaz T, Viaene J, Wigier M (eds) Innovation in small firms and dynamics of local development. Scholar Publishing House, Warsaw, pp 13–31

Noronha Vaz T (2011) The design of industrial models: addressing cooperative behaviours, innovation and public policy. In: Desai S, Nijkamp P, Stough R (eds) New directions in regional economic development: the role of entrepreneurship theory and methods, practice and policy. Edward Elgar Publishing, Cheltenham

Noronha, T and Vaz, E. (2015) Framing urban habitats: the small and medium towns in the peripheries, Habitat International, Special Issue: Exploratory Spatial Analysis of Urban Habitats Volume 45, Part 2, January 2015, Pages 147–155

Noronha Vaz T, Cesário M (2004) Territorial systems in the rural areas of the European Union. New MEDIT Journal, n° 4/2004, Ediziones Dedalo, ISSN: 1594-5685, Vol. III, n°4: 11–22

Noronha Vaz T, Cesário M (2008) Driving Forces for Innovation: Are They Measurable? Int J Foresight Innov Policy 4(1-2):30–50

Noronha Vaz T, Nijkamp P (2009a) Knowledge and innovation: the strings between global and local dimensions of sustainable growth. Entrep Reg Dev 21(4):441–455

Noronha Vaz T, Nijkamp P (2009b) Multitasking in the rural world: technological change and sustainability. Int J Agric Res Govern Ecol 8(2):111–129

Noronha Vaz T, Barbosa A, Cesário M, Guerreiro A (2003) Regional attractibility to business. An empirical application to southern European regions. New MEDIT J, University of Bologna, Itália., ISSN: 1594-5685, n° 3/2003:52–57

Noronha Vaz T, Leeuwen ES, Nijkamp P (2013) Lessons from successful small towns. In: Noronha Vaz T, Leeuwen ES, Nijkamp P (eds) Small towns in the rural world. Edward Elgar Publishing, Cheltenham, p 367–371 and 367–373

OECD Statistics and World Economic Outlook data, 2013

Piketty T, Goldhammer A (2015) The economics of inequality. Harvard University Press, London

Pinto H, Noronha T, Faustino C (2015) Knowledge and cooperation determinants of innovation networks: a mixed-methods approach to the case of Portugal. J Technol Manage Innov 10(1): 83–102

PORDATA and Eurostat and National Institutes of Statistics

Quévit M (1995) The regional impact of the internal market: a comparative analysis of traditional industrial regions and lagging regions. In: Hardy S, Hart M, Albrechts L, Katos A (eds) An enlarged Europe. Regions in competition? Regional policy and development. Jessica Kingsley Publishers and Regional Studies Association, London, pp 55–69

Romer PM (1986) Increasing returns and long-run growth. J Polit Econ 94-1986:1002–1037

Solow RM (1956) A contribution to the theory of economic growth. Q J Econ 70:65–94

Vaz E (2016) The future of landscapes and habitats: The regional science contribution to the understanding of geographical space. Habitat Int 51:70–78
Vaz E, Noronha T, Nijkamp P (2013) An exploratory landscape metrics approach for agricultural sustainability. Agroecol Sustain Food Syst 38(1):92–108
Vaz E, Cabral P, Caetano M, Nijkamp P, Painho M (2012) Urban heritage endangerment at the interface of future cities and past heritage: A spatial vulnerability assessment. Habitat Int 36(2): 287–294
Vaz E, de Noronha Vaz T, Galindo PV, Nijkamp P (2014a) Modelling innovation support systems for regional development–analysis of cluster structures in innovation in Portugal. Entrepreneurship & Regional Development 26(1–2):23–46
Vaz E, De Noronha T, Nijkamp P (2014b) Exploratory landscape metrics for agricultural sustainability. Agroecol Sustain Food Syst 38(1):92–108
Vaz E, Nijkamp P (2015) Gravitational forces in the spatial impacts of urban sprawl: An investigation of the region of Veneto, Italy. Habitat Int 45:99–105
Zamora-Kapoor A, Coller X (2014) The effects of the crisis: why Southern Europe? Am Behav Sci 58(12):1511–1516

Chapter 2
Regional Opportunities in Southern Europe

Abstract Southern Europe integrates a significant set of hinterlands and peri-urban regions that define an extensive rural reality rooted in tradition, specialized economy, and local identities. Nevertheless, a concerning amount of loss of regional identity has been found by means of change from this rural identity that carries strong historical roots. The heterogeneity of southern Europe's region questions the opportunities of hospitality sectors as a potential exploration for new determinants of economic growth in these regions. By considering the investment in skills, infrastructure, and regional growth, it is of utmost importance to consolidate the identity of the rural world in southern Europe, with key drivers related to an integrative approach to governance system. The chapter goes on to discuss these approaches, and how they can pave a way toward a more resilient and functional southern Europe considering the spatial and sustainability dimensions of lagging regions and the importance of rural and urban interaction.

Keywords Rurality · Industrial organization · Firms · Local governance · Southern Europe

2.1 A New Scenario for the Rural World

The increasing population in urban areas is asymmetric compared to the vast rural land use we have through the world. The urbanization process is characterized not only by the phenomenon of intensification of built-up land in metropolitan regions and cities or their continuous concentration and sprawl worldwide. Indeed, the dichotomy of urban versus rural has been gradually disappearing, leaving a footprint of economic, ecological, and social development of cities scattered across rural regions, partially changing much of their traditional function. That does not mean that traditional values have totally disappeared but, rather, that they have an increasing number of activities that are likely found in urban economies. As such, rural areas can no longer be understood only as the natural landscape of agricultural activities: The concept of rurality changed over time (Noronha Vaz et al. 2006), integrating some aspects of urbanization. In addition, rural areas increasingly host

E. Vaz, T. de Noronha, *Sustainable Development in Southern Europe*,
https://doi.org/10.1007/978-3-662-62177-6_2

the leisure industry, by means of endogenous tourism experiences, tailored in the historical and the rural world (Vaz et al. 2011a; Vaz and Agapito 2020). As a result, rural areas are currently at an interface of two sets of forces: the local and regional development that occurs in nearby areas and the globalization which may affect in various ways the sustainability of their fragile production systems.

After the end of the Second World War, a new paradigm began marking much of the western countryside. The search for economic conversion quickly intensified agricultural activities based on a model of productive, intensive, mechanized, and stable specialization despite the plummeting in agricultural prices. At the same time, European agriculture was driven by the priorities set by the CAP (Common Agricultural Policy), with the primary objective of stabilizing farmers' incomes (Daugbjerg 1999). This strongly affected countries such as Greece and Portugal. Simultaneously, the United States explored agribusiness and agricultural production reaching high levels of national outcomes, growing the agri-food sector and its exports, as well as generating a complex and ambiguous concept of rurality.

In Europe, the understanding of the rural context cannot be disassociated from the Common Agricultural Policy. Since its foundation, CAP has generated deep structural changes such as (1) land use reforms, (2) efficiency of payment systems, (3) development of cooperation structures, (4) reorganization of distribution, and (5) logistics and marketing channels for promotion of agri-food sectors. Most of such changes occurred in the rural peripheries with impacts on land use and a changing perception of the rural world. This proved to be beneficial for southern Europe supporting its modernization in the urban as well as the rural nexus.

Hence, it is not surprising that in the 1970s, the guaranteed agricultural prices had strengthened the success of the European agricultural sector and provided a continuous rural development towards new social living standards – some partly urbanized in much of the more developed European countryside of central and northern European countries (Richardson 2000). Such instruments, independently from those incentives for regional development included in the EU fund for regional development, FEDER, have generated severe regional imbalances throughout Europe with an increasing north versus south gap (Cappelen et al. 1999).

Studies confirm that the evolution of the concept of rurality is becoming increasingly complex and diverse. For example, using samples for different countries, Gülümser and others (2008) select rural indicators to apply a multidimensional classification technique, comparing the different levels of rurality among the Member States of the EU and Turkey. Even considering only five factors as major drivers of rural areas (population, employment, income, education, land use and energy use), the results show interesting differences in the values achieved by each country. Although the authors have not managed to include in the study important factors such as innovation and exports or imports, it was possible to separate the indicators which contributed positively to the rural world (such as demographics) of those with a negative impact (higher education and industrialization). The authors confirm that northern and western European countries are not intrinsically rural. Contrastingly, southern and eastern Europe confirm our analyses of the previous chapter.

The conclusion is fundamental to our discussion defining a first aspect, central to the study of southern Europe: its rural identity, now more complex and diverse. South Europe holds a strong and diverse rural identity, an intrinsic part of its cities and regions.

By exerting a profound impact on the spatial characteristics of economic activities, globalization affects the location decisions of corporations, as well as their organization and logistics. Thus, sustainable development of the rural world requires adequate environmental conditions, which can maintain the rural characteristics and simultaneously allow business expansion. These are difficult trade-offs between agricultural interest and the importance of rural activities, the new role of recreation and leisure in the emergence of small-scale business activities that match the rural landscape, and the necessity of preserving an environmental sustainability.

In this context, how should sustainable development of rural southern Europe be achieved? Four important issues deserve attention to explore these possibilities for sustainable development of southern Europe's rural regions: (1) the specific features of rural areas and their types of economic activity (Vaz et al. 2012; Vaz 2016), (2) the effects of the predominance of small companies with their vital links to external environment, (3) the development of smart regions through technological integration in combination with the emergence of new models of industrial production and organization (Vaz et al. 2017; Samora-Arvela et al. 2018; de Noronha and Vaz 2020), and (4) the opportunities from various types of public policies to promote development in the area based on the increase of the already existing knowledge base frequently with the help of regional university and other learning institutions.

The main economic activity in rural areas of southern Europe is based on traditional sectors such as agriculture, gastronomy, arts and crafts, and tourism, which absorb little technological change and therefore are more vulnerable to the competition within a global world. Local development of peripheral areas in countries, including southern Europe, is mainly determined by small companies with low investment, lack of access to knowledge, and limited capacity to absorb new technologies (Vaz et al. 2011b, 2015).

At the same time, the traditional location of industrial activities becomes increasingly oriented towards target modes of production, prevailing moving towards those specialized activities able to respond flexibly to new market demands, locally as well as globally. The likely result is the increase of trade flows for goods and services as well as for information and knowledge, tending to intensify interactions among regions that best can partake in novel developments. Figure 2.1 reports the percentage of young demographics not working or studying across Europe and illustrates the major weakness of southern Europe. In a context where knowledge and advanced skills are the major factors to speed up innovation and progress, the greatest concerns for the future of southern Europe are classical: population aging, lack of skills, entrepreneurship, and investment. This is the fact to be noted before any other strategic agenda related to the industry (equipment or accessibility construction) or services (tourism or natural resource exploitation) promoting growth (Vaz et al. 2013).

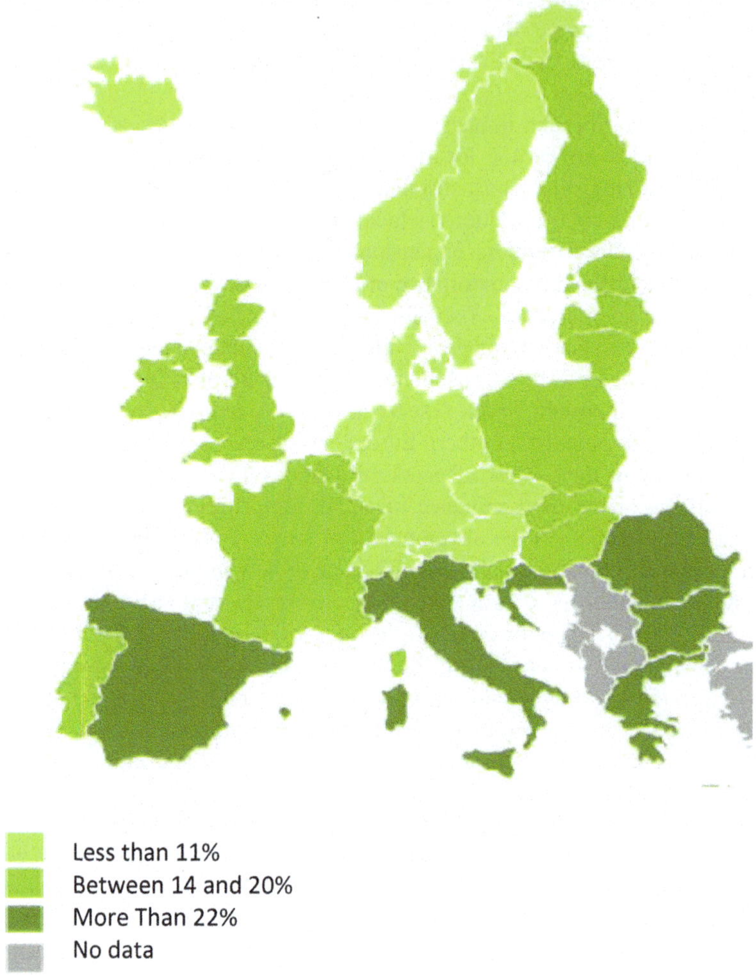

Fig. 2.1 Percentage of youngsters not working or studying in Europe, by country. (Source: Eurostat)

In the developed world, agriculture is a wide range activity with a close connection to the food industry, affecting in many ways the land use of the countryside. Acquisitions by large corporations or imposed coordination forms by means of contractual relationships have altered the ownership structure of properties. In the case of southern Europe, however, and facing a more recent set of financial supports, several associative practices have been promoted by the European Commission to help solve some of the natural growing tensions within vertical relationships between farmers and distribution in the agri-food sector.

Weiss (1974) suggested how market power has consequences for industrial organization including agriculture, and Cotterill and Iton (1993) studied the structure-performance relationships within the food systems, concluding that concentration is one of the determinants of business profitability in the sector. These arguments had a clear impact on the development or rural areas across Europe, but the southern regions stayed mostly apart of the political efforts at the time.

For a long period, much of the efficiency of the food supply chains across the world were a result from vertical coordination in the distribution segments, determined by stressing prices at producers' level and imposing tensions in payment conditions in search of more advantages for the distributors.

In Europe, the agricultural and food policies have clearly biased the revenues from agriculture with a strong impact for many years in the rural activities. When in the presence of a powerful market able to influence commercial channels at the global level, decreases in prices could impose serious and unexpected restrictions to farmers, consequently decreasing revenues in the rural areas and impacting its already slow development process of many towns, particularly the small towns and localities. Authors, such as Gadde (2004), have advanced, thus, with the existence of two distinct productive forms, the global and the local, mainly characterized by the structure of the marketing channels they are affected to and the proximity of such structures to the consumers.

In the case of the European market for several key branches of the food sector like cereals, dairy products, or wine, for many years, the CAP applied a very complex price support system that influenced farmers in a way that was almost totally out of their control and accentuating the relationship of European agriculture to industrial mass production and global markets. In a much less interventionist political context, also the American primary sector has the same industrial key determinants. Boyer and Durand (1998) have confirmed that such agri-food systems lack efficiency; thus they not only prevent technical change and upgrading but also take to a disruptive process based on local markets. Further scientific confirmation was given by Hinrichs (2000). Indeed, local identity allied to labor specialization is catching an increasing number of consumers, based on a better guarantee of quality and traceability.

We could advance that this context of a flexible productive process offers windows of opportunities and may offer ways out to growing consumers' concerns related to unrestrained industrialization of foods, upon human and animal health. Consciousness emphasizes the basis of what we consider to be new drivers for agricultural production: (1) food safety, (2) long-term impact on the environmental balance, (3) better wealth redistribution, (4) ethical animal use, and, finally, (5) fair trade on a world scale. These trends will certainly influence the upcoming development in southern Europe (Covas 2007), hopefully positively.

The acceptance of other successful productive forms based upon segmentation and networking suggests a solid argument in favor of local production and proximity circuits, thereby reinforcing new food systems (Parrot et al. 2002). Nowadays such forms are based upon labor specialization and combine long-term learning processes, thereby facilitating technological innovation to break through in less favored regions or locals, in a process which may be crucial for the sustainability of southern Europe.

In fact, flexible production can reach extreme situations of unorganized and very dispersed forms: a significant number of cases were observed and identified as "alternative agri-food initiatives" (AFIs). Identified as emerging structures of community-supported agriculture, they include farmers' markets, urban agriculture, and regional food labels, all novel actions involving consumers to traditional farmers and production (Noronha Vaz and Nijkamp 2008a, b).

It is important to accept the driving role and contribution to social sustainability that those small rural areas have in the development of urbanized areas. There is a complementarity to be accessed in South America, Africa, and India, where urbanization also depends on how rural small towns support a lifestyle able to provide for social security. Such inputs ease growth paths for development of many countries in such continents (Hayami 2007). The same has also been considered for the case of southern Europe, where small- and medium-sized towns become a very important core for the dynamics of change of the whole region (Noronha Vaz et al. 2013). Contrarily, in Europe (northern and central areas) and North America, rural development has a much-restricted perception, and links are not the same. Rural areas provide better conditions for an agrobusiness model to occur, the social responsibility eventually failing.

Finally, a very important argument that must be taken in consideration is that although the evidence shows the existence of a diversified rural world, a common ground is rural identity.

Charged by the historical past, rural identity is frequently identified in southern Europe in a specific profile of traditional values, significant amount of tacit knowledge, and much reluctance to change. Moreover, in a globalized Europe, an opportune question is whether rural identities are tending to merge into a global and unique concept or, on the contrary, each rural will maintain its local characteristics as positive contributions for regional and national competitive performances. Figure 2.2, about the performance of intermediate and predominantly rural regions across Europe in 2013, can offer some insights to the questions demonstrating that among rural Europe, different trends must be considered and that those that most affect southern Europe are predominantly or almost depleting.

Food production as a development regional strategy for southern Europe should be defined as a priority (Vaz et al. 2015).

However, the way to follow is tangled. Firstly, considering the need for internal economies to get size or scope advantages in the production process, the role of competition and technical innovation becomes a necessary tool. If diseconomies start, the disaggregation of the productive system occurs in search of innovation for which small firms are the best agents if able to fast adapt the technical specificities of diversification: specific knowledge, skills, and investment or entrepreneurship are a must, and political support at the regional level and tailor-made incentives are the engine for structural change in southern European regions.

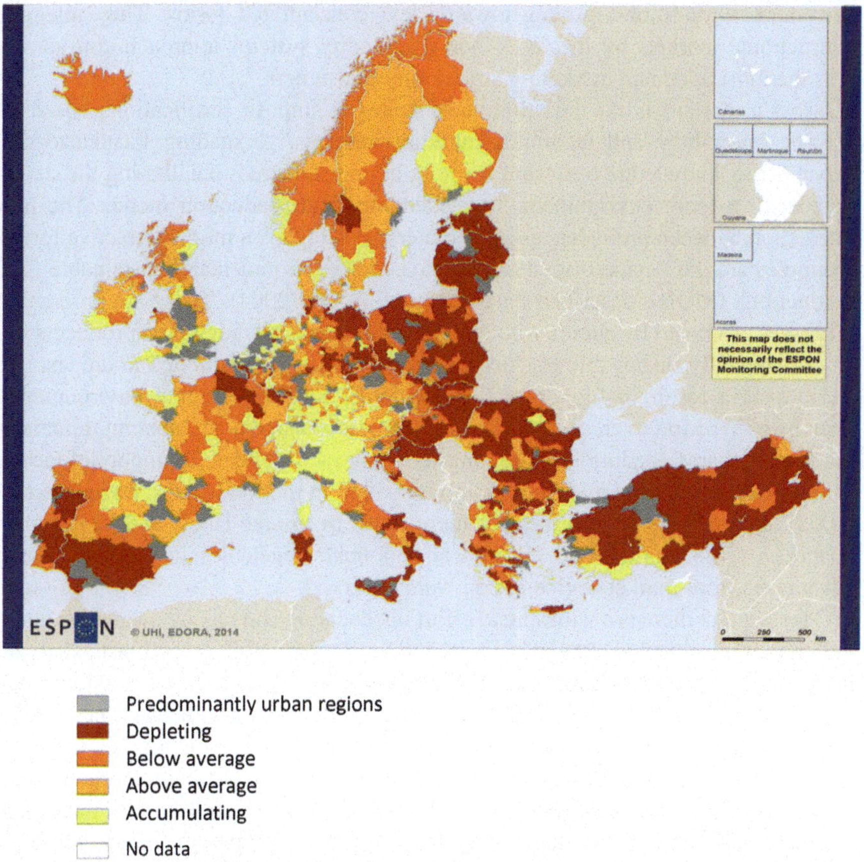

Fig. 2.2 Performance of intermediate and predominantly rural regions, 2013. (Source: ESPON 2014)

Local competition can only be understood if the culture of the firm and the level of embeddedness are accessed by those links that farms and firms in rural environments develop with other institutions. These links tend to be instable and can occur with external agents. If the segmentation process is frequent and dominates the productive processes, more conflicts and contracting tensions are to be expected (Noronha Vaz et al. 2008).

How to surpass such hindrances depends on the organizational forms of the productive systems and how they can find suitable forms for efficient coexistence. Among other things, innovation and regulation provide insights for good results mainly in cooperation and networking. There are essays describing success cases such as Spanish olive oil (Mili 2008), Belgium beer (Avermaete and Vandermosten 2008), Italian wine (Gatti 2008), or Brazilian coffee (Urban and Vaz 2008), cases of market segments earlier focused on mass production industries but today adapting to new market conditions moving towards less concentrated forms. They integrate international markets by means of non-commodity systems approach considering it as the most adequate marketing strategy for consumers.

Most important is the fact that such strategies impose certifications based on specific know-how and technological improvements, persuading local actors to expand their knowledge basis and learning attitudes, thereby facilitating the development of a more environmentally sustainable food production model. The new interaction between producers and consumers is taking even more interactive forms. Examples are the "market-oriented initiatives for environmentally sustainable food production" (MOIs), described by Buller and Morris (2004). These are production processes promoted by clients who increase demands once they know producers use environmentally sustainable practices. Private labeling has exploded in the last two decades as a result of higher-quality standards, food safety, and environmental responsibility. Moreover, and in terms of business management, labeling influences the supply chains, leading to more long-term contracts generating important factors for introduction of new rules and more innovation in the food chain. Bazoche et al. (2005) analyzed the commitment of these kinds of private labels and their consequences on the market prices for meat and fresh vegetables in France; several advantages from market segmentation were observed.

The fact that there is a widespread effort in educating and raising awareness about the dietary priorities of population in general to favor healthier food brings advantages to local production processes, not only because consumers are ready to change their consumption conducts (Guptill and Wilkins 2002) but also because new praxis alerts for the ethical dimension and calls for those best practices in terms of environmental sustainability. Gatti (2008) points out the wine sector in Emilia-Romagna, Italy, and confirms that wine, one of the most well-known traditional products, will get more and better markets if its productive process can be linked to environmentally friendly production, thus increasing sustainable impacts.

Thus, it seems important to reinforce that in an evolutionary market situation, as earlier defined, the cohabitation of the two different productive models, mass production and segmentation, may take place, eventually, even at the same time and place (Vaz 2008). Most probably, in the complexity of those generated interrelationships, business organization becomes inevitable, predominantly, in routines of shared information in order to reduce uncertainty and risk – product traceability occurs and becomes the essential solution to unify and identify agents along the food chain, or public-private partnerships emerge as regulators of food safety (Galizzi and Venturini 1999).

As new forms of industrial organization are becoming more important (an example is the adoption of new technologies that can result in modular and flexible forms of production emerging in large production units, previously designed in economies of scale and integrated processes), the sites are competing with each other in terms of their ability to offer companies the best conditions and opportunities to decentralize production processes and management.

Finally, the different types of public support and the accompanying policies required to support the transition to a new type of rural sustainable world need to be distinguished and perceived. For example, the ability to learn is an important determinant of innovation, so that the technology support to address the learning environment and process for domestic and foreign firms must be considered.

In the case of Europe, there are many niche markets where traditional forms of production are still dominant and for which consumers look with interest – the wide variety of geography may stimulate southern European territorial identity and promotes locals, always creating new expectations for products and services with typical or specific attributes. The traditional production methods are a precious legacy of the past and as such, being a mirror of the historical component of sites, may have some chances of survival. These sectors generally low-tech have a place, even if sometimes modest, in our contemporary society and particularly in the rural periphery, also as basins of knowledge and valuable information or as spreader of creative production processes over time.

Such environments, essentially established by structures of small businesses, put up the challenge of bringing technological and organizational innovation to places where the common industrial model has difficulties in being applied.

Belonging to the low-tech sectors, most firms that produce traditionally (foods or other items, it does not really matter) are compelled to achieve their goals and produce facing the global limitations of a growing competition. So, innovativeness is the major tool able to guarantee for firms' revenues and subsequent local prosperity. For rural small businesses, often located far from the key commercial circuits and, in general, without sufficient marketing skills, market share quotas are basically defined by proximity.

Although integrating the several business requisites, such as technological training, fast fine-tuning to modern business environments, and knowledge acquired in a historical context, joint actions and network interdependencies are a necessary requirement to get new meanings for rural firms. Research, selection and codification, transformation and quality control, and other procedures need to be an inherent part of all the rural processes whatever the products may be.

Today's information systems facilitate the creation of specialized networks, thereby easing sustainable platforms for regional development. Marsden and Smith (2005) confirm that such kind of networks promotes and shapes knowledge facilitating the willingness to innovate. By using a combination of fragmentation, specialization, and quality building strategies, those firms may encourage sustainable development in rural areas in general and in southern Europe in particular.

As a final note, it is necessary to emphasize the multifunctional character of the rural world, well described by Peterson et al. (2002), van Huylenbroeck and Durand (2004), and Noronha Vaz and Nijkamp (2009), which allows the entrance of other significant opportunities and different lifestyles to generate knowledge and local incomes; other activities are emerging that may bring new life to the rural scene.

2.2 Governance Systems

Sustainable development: So far, arguments used to explain prosperity of the rural world have had a functional economic nature. However, because of the emergence of the ecological awareness, the standpoint of sustainability as inherent to its development process has long been accepted (van den Bergh 1996). Thus, land use and land planning became major contributors to ensure the sustainability of rural areas and help in environmental management and agricultural land use policy (Oltmer et al. 2008).

Today there is a general trend towards the loss of natural land in favor of cropland by means of a slow but steady conversion of natural areas into areas for agricultural, urban, or industrial uses.

Multidisciplinary approaches: There is though the need for an efficient use of the natural resources which pushes the concept of rural development towards sustainability, a better context to analyze issues such as land use, quality, and tolerance of ecosystems or biodiversity. Pohoryles (2007) confirms that an initial policy approach evolved to a global dimension, not limited to specific regions and as a horizontal concern. In the spirit of this debate, land use change has recently developed in a new focal point of interest for both scientists and policy makers. Issues such as deforestation (Chomitz and Gray 1996), soil rehabilitation (Beinat and Nijkamp 1998, 1999; Nijkamp 2000), and, in particular, urban renewal (Finco and Nijkamp 2000) outstretched. Also, a new multidisciplinary approach for such issues including economic, demographic, technological, or physical nature analyses emerged, justifying more than before the need for regional or local level assessments (Giaoutzi and Nijkamp 1994).

Regional institutional environments: So far, we considered several market mechanisms that could bring change to the rural world. However most of the ability to promote them relies on decisions, adjustments, and regulations that originate in the governance systems, which go beyond the private sphere and become public political issues. How the institutional settings can work out market trends, such as those earlier mentioned creating or promoting mechanisms for regional and local innovation systems, is one of the hardest challenges for southern Europe. As explained by Neto et al. (2008), in the rural areas, the potential for the private activities and entrepreneurship frequently depends on public initiatives and projects which are defined in the scope of regional policies, although frequently promoted by the European framing; regulations and local organizations should form the best institutional environments for adjustment also involving stockholders, whenever possible. They act and participate actively in policy making and decentralization processes such as land use planning.

Yet, on this regard, much more could be debated in what concerns the political choice, the political power, the game of influences, and the degree of awareness of civil population in general, a major feature to design the complete governance system at decentralized level.

Research and development: So far, the impact of knowledge production in southern Europe has not yet been the focus of our attention, but this is a central question for any peripheral region. Whether or not private or public funds should be addressed as investment in R&D depends mostly on how expressive and regular the spillover effects are. Still today this topic remains controversial.

The theoretical approach is led by the neoclassical growth theory; Solow (1994) suggests that the growth path of any given country or region predicts productivity increases because of higher amount of capital each worker is set to activate. However, as capital per worker increases, its marginal productivity decreases, due to the diminishing returns of capital and, with it, the scope for further increases in the capital-labor ratio. So, the theory advocates that increasing investment in core regions becomes progressively inefficient and investments in the periphery a more attractive option leading to convergence among regions.

Despite these considerations, Romer (1986) and Lucas (1988) had long emphasized technology as an endogenous, major factor affecting economic growth and increasing returns to investment may frequently turn to divergence of regions. Meanwhile, many empirical cases have demonstrated that core regions rather than the periphery investment tend to return better and faster results in terms of innovativeness and positive multiplier effects.

In the case of southern Europe, one of the main concerns is the fact that R&D activities are costly and require a critical mass before being capable of generating technological progress. Traditionally, these regions have required a straightforward scientific and technological regional strategy and the firms should create links with other locations. This difficulty associated with budgetary restrictions often results in insufficient allocation of resources devoted to R&D, many concluding that R&D expenditure may have little real effects upon the economic progress of these regions.

Furthermore, increasing returns to investment on R&D activities are to be expected. Scherer (1982) underlined the positive economies of scale and scope derived concentration of such activities to which Fagerberg (1988) confirmed that R&D investment returns benefit from strong cumulative effects, maximizing the positive externalities from the agglomeration of R&D activities as later referredby Audretsch and Feldman (1996) or Verspagen (1997).

A second factor preventing peripheral regions from investing in R&D which was firstly mentioned by Storper and Harrison (1991) is the fact that technological innovation permits new ideas generated at a zero marginal cost that are not appropriable. Also confirmed by Harari (1995), technological developments tend to be mobile, and although firms try to appropriate the returns of their research initiatives, the most common forms of appropriation (patents, lead time) are regarded as highly imperfect; thus technological novelties can easily spill over to the rest of the community.

Much attention should be given to the transaction cost approach, which proposes that there is no one sole solution for the above-described dilemma, much depending on the type of transactions taking place and the power of involved agents. The goal is the pursuit of efficiency under the assumption that the market and its atomistic organization consolidates all the relations into transactions ruled by the rationality and the opportunism of agents (Williamson 1975 and 1985 in Taylor 1995):

1. If transactions involve great uncertainty, they occur in hierarchically organized contexts to saving costs in marketing transactions but, at the same time, using more bureaucracy.
2. But flexible specialization and flexible accumulation is also possible when replacing hierarchies in localized market transactions. Storper and Harrison (1991) draw attention to the structure of the supply chains where the number of suppliers is different from the number of customers. The role of power or unequal power relationships within and between firms even influences patterns of geographical industrialization, speeding regional development.

References

Avermaete T, Vandermosten G (2008) Traditional Belgian beers in a global market economy. In: Vaz MTN, Nijkamp P, Rastoin JL (eds) Traditional food production facing sustainability: a European challenge. Ashgate, Aldershot

Audretsch DB, Feldman MP (1996) Innovative clusters and the industry life cycle. Rev Ind Organ 11(2):253–273

Bazoche P, Giraud-Héraud E, Soler L-G (2005) Premium private labels, supply contracts, market segmentation, and spot prices. J Agric Food Indus Org 3(1). https://doi.org/10.2202/1542-0485.1087

Beinat E, Nijkamp P (1998) Environmental rehabilitation: efficiency and effectiveness in soil remediation. Stud Reg Urban Plann 6:83–101

Beinat E, Nijkamp P (eds) (1999) Multicriteria analysis for land-use management. Kluwer, Dordrecht

Boyer R, Durand JP (1998) L'après Fordisme, La Découvert and Syros Alternatives Economiques, Paris

Buller H, Morris C (2004) Growing goods: the market, the state, and sustainable food production. Environ Plann A 36(6):1065–1084

Cappelen A, Fagerberg J, Verspagen B (1999) Lack of regional convergence. In: Fagerberg J, Guerrieri P, Verspagen B (eds) The economic challenge for Europe: adapting to innovation-based growth. Edward Elgar, Aldershot, pp 228–237

Chomitz KM, Gray DA (1996) Roads, land use, and deforestation: a spatial model applied to belize. World Bank Econ Rev 10(3):487–512

Cotterill RW, Iton CW (1993) A PIMS analysis of the structure profit relationship in food manufacturing. In: Cotterill RW (ed) Competitive strategy analysis in the food system. Westview Press, Oxford, pp 25–44

Covas A (2007) Ruralidades II Agricultura Multifuncional e Desenvolvimento Rural. Universidade do Algarve, Faro

Daugbjerg C (1999) Reforming the CAP: policy networks and broader institutional structures. J Common Mark Stud 37(3):407–428

de Noronha T, Vaz E (2020) Theoretical foundations in support of small and medium towns. Sustainability 12(13):5312

Fagerberg J (1988) Why growth rates differ. In: Dosi G, Freeman C, Nelson R, Silverberg G, Soete L (eds) Technical change and economic theory. Pinter, London, pp 432–457

Finco A, Nijkamp P (2000) Towards a sustainable future of cities in Europe. In: Stillwell J, Scholten H (eds) Land use simulation for Europe, vol 2001. Kluwer, Dordrecht, pp 173–192

Gadde L-E (2004) Activity coordination and resource combining in distribution networks – implications for relationship involvement and the relationship atmosphere. J Mark Manage 20(1-2): 157–184

Galizzi G, Venturini L (1999) Vertical relationships and coordination in the food system. Physica-Verlag, Heidelberg

Gatti S (2008) Protected designation of origin, sustainable development and international policies: a survey of DOC wines from Emilia-Romagna. In: Vaz MTN, Nijkamp P, Rastoin JL (eds) Traditional food production facing sustainability: a European challenge. Ashgate, Aldershot

Giaoutzi M, Nijkamp P (1994) Decision support models for regional sustainable development. Ashgate, Aldershot

Gülümser AA, Baycan-Levent T, Nijkamp P (2008) A comparative analysis of rurality at the EU Level and Turkey. In: Vaz MTN, Nijkamp P, Rastoin JL (eds) Traditional food production facing sustainability: a European challenge. Ashgate, Aldershot

Guptill A, Wilkins JL (2002) Buying into the food system: trends in food retailing in the US and implications for local foods. Agric Human Values 19(1):39–51

Harari N (1995) Appropriability of technical innovation an empirical analysis. Res Policy 24(6): 981–999

Hayami Y (2007) From the megalopolis-centred system to the rural-urban balanced system for the sustainability of developing economies. Keynote address at the Science Council of Japan Conference on Sustainability, 7-8 September, The Foundation for Advanced Studies in International Development

Hinrichs CC (2000) Embeddedness and local food systems: notes on two types of direct agricultural market. J Rural Stud 16(3):295–303

Lucas RE (1988) On the mechanics of economic development. J Monet Econ 22-1988:3–42

Marsden T, Smith E (2005) Ecological entrepreneurship: sustainable development in local communities through quality food production and local branding. Geoforum 36(4):440–451

Mili S (2008) Market dynamics and policy reforms in the olive oil sector: a European perspective. In: Vaz MTN, Nijkamp P, Rastoin JL (eds) Traditional food production facing sustainability: a European challenge. Ashgate, Aldershot

Neto P, Couto JA, Natário MM (2008) Governance and the determinants of local economic development. In: Vaz MTN, Nijkamp P, Rastoin JL (eds) Traditional food production facing sustainability: a European challenge. Ashgate, Aldershot

Nijkamp P (2000) Critical success factors for soil remediation policy. J Environ Policy Law (1):81–98

Noronha Vaz T (2008) The design of industrial models. In: Desai S, Stough R, Nijkamp P (eds) Entrepreneurship and innovation: a spatial perspective. Edward Elgar, Cheltenham

Noronha Vaz T, Nijkamp P (2008a) The complex force field of traditional food systems: setting the scene. In: Vaz MTN, Nijkamp P, Rastoin JL (eds) Traditional food production facing sustainability: a European challenge. Ashgate, Aldershot

Noronha Vaz T, Nijkamp P (2008b) Large-scale production and market segmentation: an uneasy relationship. In: Vaz MTN, Nijkamp P, Rastoin JL (eds) Traditional food production facing sustainability: a European challenge. Ashgate, Aldershot

Noronha Vaz T, Nijkamp P (2009) Multitasking in the rural world: technological change and sustainability. Int J Agric Resour Govern Ecol 8(2):111–129

Noronha Vaz T, Morgan E, Nijkamp P (eds) (2006) The new European rurality: strategies for small business. Routledge, London

Noronha Vaz T, Nijkamp P, Rastoin JL (eds) (2008) Traditional food production and rural sustainable development: a European challenge. Routledge, London

Noronha Vaz T, Leeuwen ES, Nijkamp P (eds) (2013) Towns in the rural world. Routledge, London

Oltmer K, Nijkamp P, Florax R, Brouwer F (2008) Sustainability and agri-environmental policy in the European Union: a meta-analytic investigation. In: Vaz MTN, Nijkamp P, Rastoin JL (eds) Traditional food production facing sustainability: a European challenge. Ashgate, Aldershot

Parrot N, Wilson N, Murdoch J (2002) Spatializing quality: regional protection and the alternative geography of food. Eur Urban Reg Stud 9(3):241–261

Peterson JM, Boisvert RN, de Gorter H (2002) Environmental policies for a multifunctional agricultural sector in open economies. Eur Rev Agric Econ 29(4):423–443

Pohoryles RJ (2007) Sustainable development, innovation and democracy. what role for the regions?. Innov Eur J Social Sci Res 2020(3):183–190. http://www.informaworld.com/smpp/title~content=t713424882~db=all~tab=issueslist~branches=20
Richardson T (2000) Discourses of rurality in EU spatial policy: the European spatial development perspective. Sociologia Ruralis 40(1):53–71
Romer PM (1986) Increasing returns and long-run growth. J Polit Econ 94(5):1002–1037
Samora-Arvela A, Vaz E, Ferrão J, Ferreira J, Panagopoulos T (2018) Diversifying Mediterranean tourism as a strategy for regional resilience enhancement. In: Resilience and regional dynamics. Springer, Cham, pp 105–127
Scherer FM (1982) Inter-industry technology flows and productivity growth. Rev Econ Stat 627–634
Solow RM (1994) Perspectives on growth theory. J Econ Perspect 8(1):45–54
Storper M, Harrison B (1991) Flexibility, hierarchy and regional development: the changing structure of industrial production systems and their forms of governance in the 1990s. Res Policy 20:407–422
Taylor M (1995) The business enterprise, power and patterns of geographical industrialisation. In: Conti S, Malecki E, Oinas P (eds) The industrial enterprise and its environment: spatial perspectives. Avebury, England, pp 99–122
Urban L, Noronha Vaz MT (2008) What is new in the production of the Brazilian coffee? Decreasing exports' risks by using organizational innovation. Economie et Societé
van den Bergh JCJM (1996) Ecological economics and sustainable development. Edward Elgar, Cheltenham
van Huylenbroeck G, Durand G (eds) (2004) Multifunctional agriculture: a new paradigm for European agriculture and rural development. Ashgate, Aldershot
Vaz E (2016) The future of landscapes and habitats: the regional science contribution to the understanding of geographical space. Habitat Int 51:70–78
Vaz E, Agapito D (2020) Recovering ancient landscapes in coastal zones for cultural tourism: a spatial analysis. In: Regional intelligence. Springer, Cham, pp 9–28
Vaz E, Nainggolan D, Nijkamp P, Painho M (2011a) Crossroads of tourism: a complex spatial systems analysis of tourism and urban sprawl in the Algarve. Int J Sustain Dev 14(3–4):225–241
Vaz E, de Noronha Vaz T, Nijkamp P (2011b) Spatial analyses for policy evaluation of the rural world: Portuguese agriculture in the last decade. In: Territorial governance. Physica-Verlag HD, pp 107–122
Vaz E, Nijkamp P, Painho M, Caetano M (2012) A multi-scenario forecast of urban change: a study on urban growth in the Algarve. Landsc Urban Plan 104(2):201–211
Vaz E, Walczynska A, Nijkamp P (2013) Regional challenges in tourist wetland systems: an integrated approach to the Ria Formosa in the Algarve, Portugal. Reg Environ Chang 13(1): 33–42
Vaz E, Painho M, Nijkamp P (2015) Linking agricultural policies with decision-making: a spatial approach. Eur Plan Stud 23(4):733–745
Vaz E, Taubenböck H, Kotha M, Arsanjani JJ (2017) Urban change in Goa, India. Habitat Int 68:24–29
Verspagen B (1997) Estimating international technology spillovers using technology flow matrices. Rev World Econ 133(2):226–248
Weiss LW (1974) The concentration-profits relationship and antitrust. In: Goldschmid HJ, Mann HM, Weston JF (eds) Industrial concentration: the new learning. Little, Brown & Company, Boston
Williamson OE (1975) Markets and hierarchies. New York, 2630

Chapter 3
Landscape and Heritage in Southern Europe

Abstract This chapter offers an integrative vision of the importance of the historical and cultural legacy of landscapes toward sustainability. Adopting a spatial-explicit approach, the chapter creates a framework for reconsidering the implementation of southern Europe's heritage within a functional role of the value of the historical and sustainable landscape. The chapter goes on addressing the issues of historical identity, and leading to consider the richness and diversity of archaeological sites within the perimeter of global sustainability, leading to a global perception of the importance of heritage preservation, but also addressing the relevance of historical identity at local level for sustainable development. Finally, the chapter addresses the impacts of regional change, particularly urban sprawl on the possible loss of territorial and landscape identity. This leads to the definition of the collapsing landscape, intrinsically at risk in southern Europe, brought by a mismatch of economic options that have led to a fragmentation of the identity of regional landscape and the current choices of preservation that have to be considered within the larger settings of Europe's sustainable growth.

Keywords Historical landscapes · Regional growth · Archaeological heritage · Southern Europe · Regional sustainability

3.1 Landscape, Land Use, and Regional Change

Regional science has extensively advanced due to the contributions of methods in geographic analysis in recent years. One of the great inputs has become the ubiquitous integration of spatially explicit tools within the traditional quantitative framework of policy and decision-making, thanks to computational advances in regional science (Fischer and Getis 2010). In this sense, challenges on the landscape and the impacts in changing regions have become possible to better address. The contribution of complex spatial modelling linked with regional science has further enabled a dialog of multidisciplinary methods that not only addresses issues within at national scale but further enables the integration of local and regional methods identifying regional asymmetries and offering more integrative approaches for a sustainable

E. Vaz, T. de Noronha, *Sustainable Development in Southern Europe*,
https://doi.org/10.1007/978-3-662-62177-6_3

future. The concept of regional intelligence arises as a method to deal with the abundance of digital data and defines novel integrations for regional sciences of the future. Such methods integrate a mixture of quantitative and qualitative techniques that embroider spatially enabled techniques that forecast and quantify current and potential future concerns. The tools offered by geographic information systems have allowed such an integration. The feasibility of understanding emergent and complex spatial behaviors has now become an increasingly important tool for validation of planning, using agent-based models, and advanced nonlinear modelling approach to interpret dynamics of anthropogenic behavior and its impact on land use, landscape, ecosystems, cities, and regions. The availability of tools that cover the dynamics of land use and land cover, promoted by the integration of remote sensing technique and large data repositories, leads to integrations within geographic information systems, articulating the spatial dimension of regional phenomena, whether from an exploratory framework for planning or for geovisualization of recurrent changes. This ubiquitous understanding leads to advances in social sciences fostered by regional and human geography, where integration of socioeconomic and demographic data leads to the potential of using spatial techniques for a rigorous assessment of the complex dynamics at regional level, this brings often accurate and optimized solutions for matching quantitative and stochastic models to assess regional change, as well as enhancing the purpose of planning through integrations of techniques found in geodesign. The clear effect of anthropogenic activity becomes thus measurable, thanks to regional data repositories both over time and space, adding on a cumulative understanding of the impacts of transitions of land and exploration of consequences on social, natural, end environmental outcomes. The current changes witnessed over the last decades have not only had an incremental effect on the environment but influenced directly the zeitgeist of regions and have established an important avenue for understanding the impacts of land from a spatial and temporal perspective. A significant contribution of understanding regional dynamics has been responded by the advancement of mapping global land cover, adding a more accurate perception of the changes in land use and the consequences from a regional perspective (de Noronha Vaz et al. 2012; Vaz and Nijkamp 2015).

The advances in instruments and multi-temporal data sources such as the GlobeLand30 project have brought remote sensing as a potential player in the forefront of the future of land use classification research, where lack of high-resolution imagery is often a costly and unavailable endeavor (Arsanjani et al. 2016; Onilude and Vaz 2020). The constant change of land use, however, should be monitored at a regional perspective, allowing to (1) understand the speed of transitions of land use, (2) detect and incorporate the typology of land use changes, and (3) simulate and forecast the impacts of these changes for the future. A great ally in the understanding of land use has been the Landsat program, successfully launched in 1976, keeping an updated and continuous repository of a flow of land use data, capable of responding to regional challenges. The enhancement of spectral resolution of Landsat 8 in 2013 has shed a long-lasting potential of monitoring land cover and land use in the future. Urban growth has been a widespread phenomenon in Europe in the last decades. It can be defined as the "physical pattern of low-density expansion of large urban areas, under market conditions, mainly into the surrounding agricultural areas" (EEA Report 2006). The exodus from agricultural areas has led to a larger amount of people living in urban areas benefiting from services and facilities of modernity. A quarter of Europe's territory has been affected by urban land use, while 80% of Europeans will be living by 2020 in urban areas, reaching in some areas up to 90% (EEA Report 2006). Despite the benefits of urban growth population in general as well as of construction of large infrastructures working as regional networks, urban growth, if not articulately and not properly planned, may peril existing cultural and environmental landscapes, particularly in southern Europe, a region facing unprecedented challenges from an economic perspective. Recent European objectives have considered the constant monitoring of urban growth as measures towards preservation within city boundaries as well as protected areas which are clearly defined. Also, academic initiatives related to geographic information sciences and remote sensing have been actively engaged in monitoring and creating consciousness on global change awareness. Global change, in fact, is a phenomenon which is confronting man's response regarding pollution, environmental sustainability, and preservation. If no measures are taken in the near future, the environment and quality of living as we know may very well become jeopardized for future generations. The importance of geographic acknowledgment of urban sprawl has become clear in the 1999 European Spatial Development Perspective, where management of cultural and natural resources was one of the key topics of discussion: "Natural and cultural heritage in the EU is endangered by economic and social modernization processes. European cultural landscapes, cities and towns, as well as a variety of natural and historic monuments are part of the European heritage. Its fostering should be an important task for modern architecture, urban and landscape planning in all regions of the EU" (ESDP 1999). Although the endangerment of natural environment is no novelty, as several different species are becoming currently extinct due to pollution and environmental change, action must be taken to allow a balanced and sustainable urban growth which avoids further threat of those species and their ecosystems (Vaz 2014; Bellout et al. 2020).

Capello (2001) suggests a twofold approach of management dynamics to cope with change: short term and long term. While the short term targets on the change of direct urban action choosing more sustainable local options, the long-term dynamic should focus on a common structural change of urban change promoting new global behaviors and technologies. As the panorama of loss regarding ecosystem is analyzed, in recent years the equation considers another aspect: cultural landscape. The abovementioned twofold approach could be an interesting solution for cultural heritage which is continuously becoming endangered. Due to the means on how urban growth has destroyed many different archaeological sites and historical buildings, on one site, the short-term perspective of changing at local level may focus stakeholders on promoting and creating cultural awareness supporting, for instance, tourist products related to monument visits while allowing the possibility to finance conservation of buildings and monuments in latent danger due to air pollution natural of city habitat. As local cultural heritage becomes manageable, long-term regional policies may reflect on undiscovered and peripheral patrimony.

These measures would benefit further academic discussions of cultural backgrounds as well as allow having in thought at the long run heritage management problems regarding urban development and expansion. The integration of these tools within different frameworks accrues additional integration for urban planners as well as policy making, manifesting the ubiquitous role of these data types. Ongoing efforts to improve the temporal resolution of these products, i.e., annual and monthly basis products, have been carried out. However, the spatial resolution of these products remains a standing challenge, because the globally covered products are generated at relatively lower spatial resolution, for example, FAO at 1 km, or GLC2000 at 1 km, or MODIS at 500 m. Furthermore, while the spatial resolution of these products is not sufficient for fine-scale studies and raises uncertainty issues, their quality and reliability still remains a big challenge. For instance, it is quite vital to have accurate maps in order to implement them in our change detection algorithms and land cover/use change simulation and definition of land policy resolutions. Because as noted by land change modelling experiences, namely, it is not sufficient to know the magnitude and type of land dynamics, e.g., expansion, shrinkage, or intensification of a specific land type, it is quite important to be accurate in the occurrences of their spatial and temporal extent.

3.2 The Importance of Heritage Landscapes

As a consequence, measures in a cultural heritage preservation context regarding urban growth forecasting may be seen in the following framework: (1) short-run, local change measures; for instance, the protection from existing urban patrimony often endangered due to air pollution – a consequence of human activity within city due to gas emissions – may be created by developing an interesting, within urban area, tourist cultural heritage offer. (2) Long-run, regional change measures promote the analysis of future urban growth leading to a sounder choice of future urban development which may as a consequence benefit directly urban planning sparing at the long-run large economic investments, for example, the recent case of Stonehenge, which due to poorly planned infrastructures 40 years ago may be on UNESCO's list of endangered world heritage sites very soon (The Times, 7th December 2007). These advantages of planning sustainability put modelling capabilities and spatial analysis in an important position for future assessments, as "Models are, perhaps, the best way of understanding the land change phenomenon and anticipate correct planning activities for sustainable cities". In spite of the uncertainty of the possibilities of preservation of this specific site, Stonehenge could very well be a lesson in future preservation and assessment of the importance of urban growth regarding cultural heritage. This is a lesson to be learned that could be used to establish new trends regarding cultural heritage endangered by urban areas. The Valley of the Kings in Egypt which shows a significant increase of urban area due to the tourist industry is an example of a cultural heritage where no active measures regarding urban growth are currently being taken. The different levels have a distinct understanding of space and make it particularly difficult to draw a combined effort to use spatial information from an ontological perspective. This is largely a result brought by human interpretation of space: from a social perspective, space is linked to place, that is, to the subjective description of memory of the region, and the narrative importance of these regions given a set of emotional values. From an economic perspective, space is the territorial definition of proximities to markets and may much better be understood when adapted as location and efficiency of location for economic growth. From a natural environment perspective, space is the subset of the environment as a whole, without considering the anthroposphere. Spatial modelling as such becomes at the interface of data availability and a subjective interpretation between the fringe of information and knowledge (Fig. 3.1). The designation of environmental change comes precisely as a result from social, economic, and natural impacts human being has exerted on the environment, taking form in the limits of carrying capacity and the possible outcomes of loss of spatially explicit landscapes and human environments.

3.3 Southern Europe: The Heritage Dimension

As a consequence of economic growth, urbanization is an increasing concern in vulnerable heritage regions. This brings a negative consequence not only on the impact of diversified landscapes but also on the consequences of fragmentation of land. The impacts at local level are paramount and unprecedented, having often a negative impact on local and regional heritage, given the vulnerability of already fragile heritage in southern Europe. In this sense, location analysis combined with local and regional knowledge can however create new tools that enable efficient decision making (Vaz 2016, 2020).

In detriment of economic growth, land use diversity of non-artificial land is decreasing. This is having a negative impact on the landscape leading to permanent loss of diversified landscapes and increasing fragmentation of land use. Regional intelligence is thus a product of the interaction of local, global, and regional knowledge, allowing the creation of spatial modelling approaches to deal within the framework of better decisions. A clear way to test regional intelligence is by means of assessing human impact on the landscape and on heritage is by measuring the variations of land use change focusing in particular on the registered changes of urban land use. In this sense, urban regions represent drivers of social and economic change, offering a clearer understanding on the impacts on the structure on the ecosystem services and diversity in urban ecosystems as a whole. The complexity of these urban regions calls for a holistic perspective where urban regions should be intrinsically diverse and presence of archaeological heritage and historical landscapes catered. The consequences of anthropogenic behavior over space, in particular population growth, have led to excessive urban sprawl and severe impacts on land use, leading to irremediable loss of heritage. Spatial models can thus capitalize on the carrying capacity of heritage, to test whether

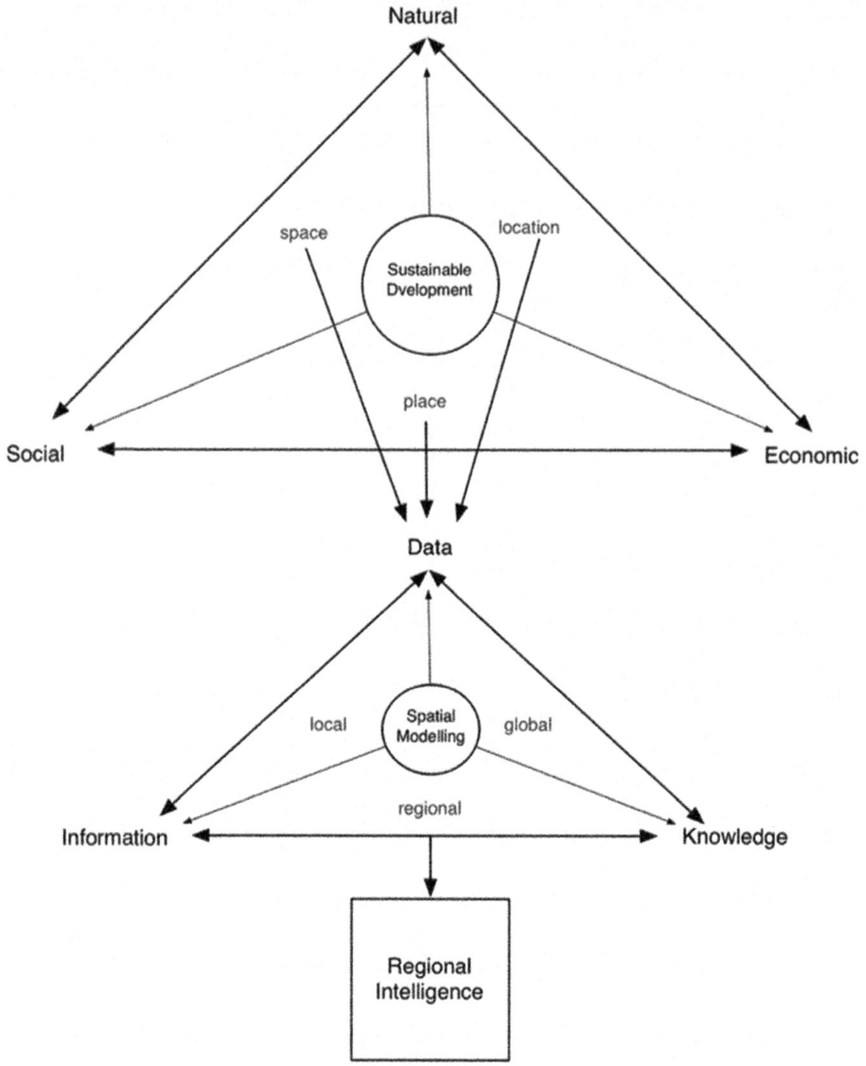

Fig. 3.1 The spatial dimension of sustainable development

regional intelligence is present. This is further registered by the fact that population increase and the urban concentration is not only creating additional pressure on the natural environment but also jeopardizing our historical ancestry, by depleting our own heritage. These landscapes share a unique and irreplaceable value that may be directly experienced, with a higher level of symbolic and cognitive value. Landscapes of the past are as such a vital part of monitoring and sustaining the landscapes of the future. The role of spatial information and geovisualization is thus to foster the role of assessing, quantifying, and identifying their risks, pressures, and shape at present. Also, the scenic values of these landscapes are important properties for sustainable tourism, permitting a diverse understanding of humankind as well as their origins and traditions. This participatory role of sustainable tourism and heritage preservation leads to a local and regional territorial identity, eventually generating a better quality of life and enhancing social responsibility for the environment in general. To evaluate and research these boundaries of spatial, economic, and social values is a fundamental role of applied regional science. Recent years have promoted the addition of spatial complexity and complexity science, where landscapes are having a key role as determinants of understanding and dealing with change. A good example is given: Cairo, one of the biggest megacities in the world, is witnessing an excessive urban growth brought by population growth and creation of new infrastructures to support tourism, economic growth, and population increase.

More than half of the world's population can be found nowadays in urban areas. The increasing growth of urban regions is leading to polycentric urban agglomerations forming a new phenomenon of urbanization throughout the world. These new geographical patterns require new planning solution to deal with the large-scale consequences of land use transitions expanding over entire regions. Within the scope of geodesign, where planning of our cities may support smarter urban areas, regional land use change should be measured to avoid strain on the carrying capacity of the environment. This is particularly true in urban and suburban regions, where population increase has been witnessed over the last 30 years. Quantifiable and analytical models may support a systemic overview of the transitions of spatial distribution, allowing foresight of changing regions. In this sense, spatial models do not offer only an integrative solution for regional analysis but a regional spatial framework which may be adopted by legislators and planners, as well as to state and local governmental entities permitting to (1) determine better land use policies and improving transportation and utility demand, (2) identify future development pressure points and areas, and (3) implement effective plans for regional development. The medium-term effect of these actions may support sustainable development at regional level, with the aim from a systemic point of view, to optimize available resources and drivers of sustainable decision-making in space. Significant contributions in the field of spatial modelling (GIS) relating land use/land cover change (LULC) have allowed for thorough empirical analysis of the consequences of land use change at different levels from a spatial modelling perspective. This is further enhanced by the availability of novel algorithms in combined remote sensing allying to the field of geographic information science, in the emergence of new techniques, tools, and methods for understanding the spatially explicit relations of cities in terms of socioeconomic and geomorphologic features within regions. The complex systems approach of space adds to the contradictory context of the urban regions: on one side, spatial modelling acts as an engine of socioeconomic growth and development prediction, but on the other, it offers a toolkit for visualization of recurring and future change and potential impacts on environmental degradation as is the surrounding natural land, heritage, and biodiversity. Techniques such as cellular automata (CA) as rule-based models given their topological and spatial characteristics are appropriate tools to incorporate spatial interaction effects and the treatment of temporal dynamics. The approach of a CA model constructs a "bottom-up" approach where the structure evolves from a Moran neighborhood interacting between neighboring cells using different types of transition rules such as multi-criteria evaluation, Markov chains, artificial neural networks, as well as pre-compiled toolboxes for urban modelling (e.g., Project Gigalopolis, MOLAND). Many studies have been undertaken and characterized by different approaches for transition rules and calibration techniques responding to different paradigms in urban and landscape sciences. This has led to the possibility of optimizing the simulation of future urban trends as well as addressing larger regions, thanks to geocomputational advances. The developing world, in particular, has seen an unprecedented expansion of urban areas and growth of urban population at such a pace that it is expected that 60% of the world's population will live in urban areas by 2030, and most of the urban growth will occur in less developed countries. Another important issue regarding urban growth is monitoring and discerning degree of sprawl to find out the amount of pressure on environment and resources, particularly at regional level. Measuring sprawl is important to plan whether the city is becoming compact or disperse, by also analyzing the rate of this move to assess the existent severity of this pressure. The physical expressions and patterns of sprawl on urban landscapes can be detected, mapped, and analyzed using remote sensing and geographical information system (GIS) technologies in conjunction with the secondary and ground truth data. This allows for a better understanding of urban sprawl in combination of spatial techniques. Although there are different approaches to measure urban sprawl, built-up areas play a key role in these different methodologies for regions. Integration of spatial accounting methods, such as deriving landscape metrics, allows for a quantification of urban sprawl and to register a better understanding of the urban form, but most of all, it may generate a better understanding of landscape fragmentation induced by anthropogenic activity. A closer analysis of the future urbanization processes of Cairo (Fig. 3.2) leads to a better understanding on how to protect fragile and important historical heritage in the future and how combinatory assessment over geographical scale must transition to a regional proxy.

Not only do regional dynamics become a concern within a metropolitan region, but the regional scale of intra-regions must be addressed to assess the relation between neighboring cities. In this sense, the distribution of this information along space allows decision-makers to understand and to visualize the consequences on the urban environment at multiple levels and may help more integrative planning avoiding collapsing regional landscapes in the future. Figure 3.3 shows the direct consequences of geovisualization at the regional level. In the case of the region of Veneto, while certain metropolitan areas share a strong connection to other metropolitan regions, it is clear that urban sprawl is predominant at the regional level, generating often unsustainable land use forms between the fringes of the hinterlands of metropolitan regions. This is a regional concern within the urban landscape of many regions throughout the world, justifying a combinatory assessment for supporting more adequate planning strategies to cope with the intra-urban growth and land use change.

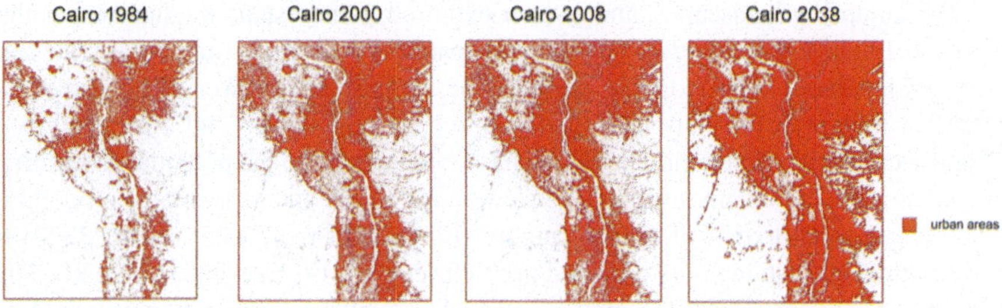

Fig. 3.2 Future urban growth in Cairo

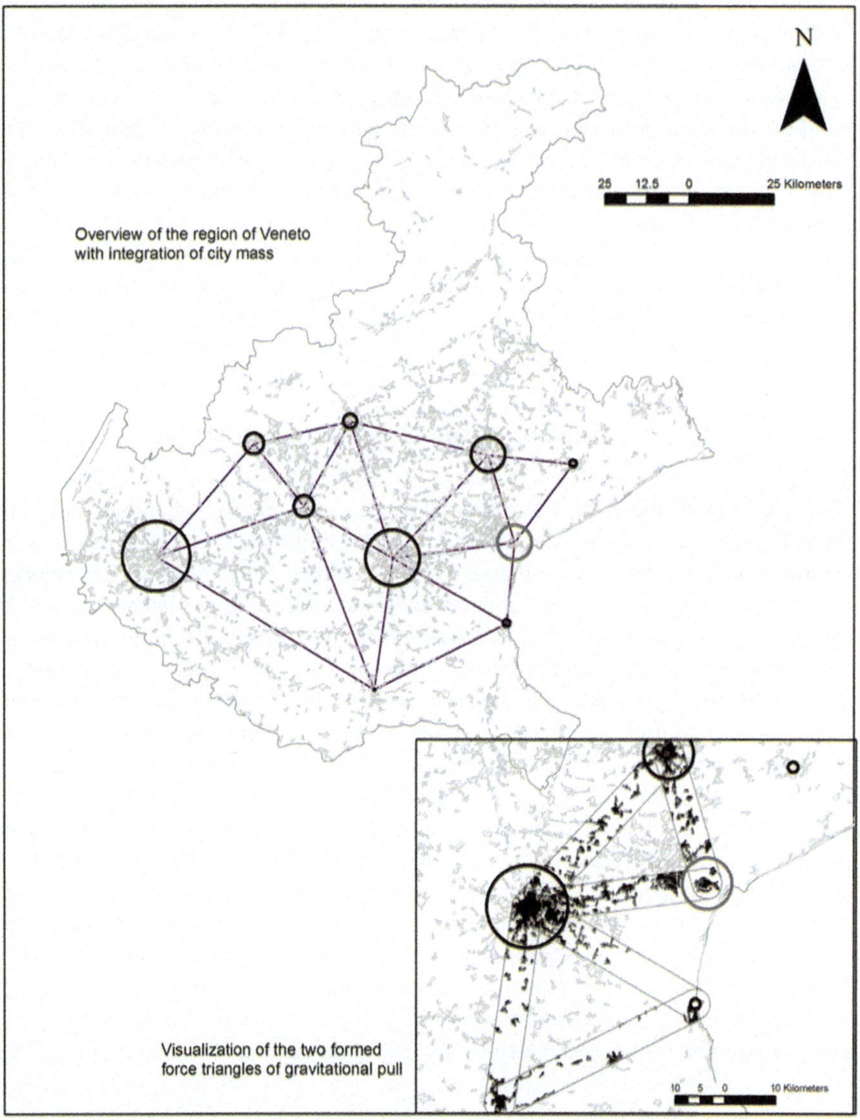

Fig. 3.3 An integrative regional assessment of Veneto's urban sprawl corridors

By collapsing landscapes, and as later explored, I understand the landscapes that have lost their territorial identity, in largest part due to human interaction with land use and increased anthropogenic activity. These landscapes, however, are part of the residual memory of society having a unique historical, social, and affective value. Signficant advances in regional sciecne have been related to the potential assessment of environmental change through geocomputation and spatio-temporal modelling, affecting a great extent of fields, ranging from health (Vaz et al. 2015, 2020) to environmental sustainability and land use (Vaz et al. 2017; Esteves et al. 2020). This "landscape collapse" is increasingly witnessed through rapid shifts in traditional land use and leads to depletion of the culture of the landscape, that is, its historical and regional territorial identity. These changes are well observed by the advances of geovisualization techniques, remote sensing, and spatial analysis and create what can be framed as regional intelligence. Witnessed at multiple scales, some of the regional evidences of the impacts of the landscape collapse are the permanent loss of material evidence found in archaeology (Monteiro et al. 2015; Vaz 2020), consequence by rapid urbanization processes, and loss of agricultural land and rural areas as well as coastal erosion, where archaeological sites have always been present. It is thus of utmost importance to consider that archaeology should have an important role in the preservation of landscape and of sustainable development. Therefore, archaeological findings must represent a parameter to measure the quality of unchanged regions and should also be adopted in a framework of regional sciences for sustainable development.

3.4 Archaeology as Part of the Regional Landscape

Quantification of information in the human sciences is not an easy process as most human sciences rely on qualitative analysis and narrative interpretation of phenomena rather than methods found in exact sciences. Some fields of the social sciences, such as sociology, geography, anthropology, and archaeology, have managed to integrate quantitative methods in the epistemological processes given their strong connection to space and geographical understanding. This quantifiable nature of social sciences has brought a convergence between mathematical and statistical methodologies to understand parts of more complex patterns and has built a bridge between quantitative and qualitative topics. This integrative approach has been interpreting the future of mankind and sustainable development. Archaeology as a human science has had an interest in quantitative and technological methodologies, supported by the possibility of quantifying material evidence. With the evolution of archaeological science and geocomputation, applications have become more pervasive, and spatial data from sites and material evidence have led to important tools for analysis, comparison, and prompting of archaeological phenomena and information. Archaeological catalogues that first started as simple archaeological registries of stratigraphic or site location interest have developed into large data containers with information which may be created, retrieved, eliminated, and changed. Hence, the basic circumstances for a database management system were established in the field of archaeology that allowed integration of information related to archaeological site phenomena. In this sense, archaeological database is not only an asset for archaeological cataloguing but also an extremely important tool for sharing information and as a consequence to use archaeological evidence to prompt for sustainable development. Information in archaeology needs, as a consequence, physical storage space, which can be stored either in manual files or in a database which may represent accurate information at the regional level. Another source of spatial data has in recent years risen: crowdsourced spatial data. This data has enabled the user to be part of the manipulation, gathering and editing of geographic information and embedding this information into a GIS. The role of the "human sensor" is one of the contributions to generating spatial information and sharing this information worldwide. This bottom-up approach is strongly linked to the advances in mobile technology, and locational technologies for the end user not only permit to pinpoint the location of archaeological and historical sites, shared by a common web GIS application such as OpenStreetMap, but, furthermore, expand on the dissemination of information generating tacit knowledge through the integration of spatial information with ancillary information resulting from social networking. Social sensitizing of historical landscapes and archaeological material evidence is thus a conciliatory dimension of the already existent GIS solutions. The location of archaeological artifacts through space may foment better planning of preservation of archaeological sites as discussed above and as foresight tools to quantify archaeological and historical presence throughout space. The traditional tools that GIS represented a decade ago enabling the user to edit, access, and visualize spatial data have now become in the web 2.0 context tools for information sharing. The human sensor has a crucial role in building on the regional importance of the broadcasting of the regional value of the landscape, but also embeds by media sharing a greater awareness of our past heritage. Archaeology has advanced within a significant transition of statistical modelling techniques due to computational advances in recent years. In this sense, current advances brought by big data analysis have made the archaeological object, such as site assessment and management, not only an objective part of heritage management but much closer to the field of regional science, understanding the ubiquitous nature of potential for economic growth and spatial decision support systems.

3.5 Spatial Solutions for Sustainable Development: A Systemic Vision

While spatial information proves to be a fundamental tool to connect landscape preservation with sustainable development, social sciences have an increasing participatory role in shaping the future. Concerning natural and historical landscapes, new integrative solutions must be built to link urban design with economic and

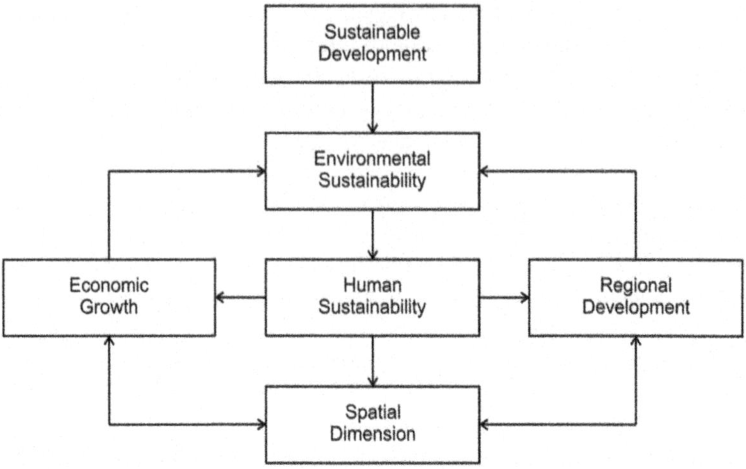

Fig. 3.4 Integration of the spatial dimension for sustainable development

environmental planning. A systemic vision of sustainable development considering spatial characteristics and the morphology of the landscape answers these challenges from a spatial perspective. In this sense human being acts as a keystone species capable of manipulating and changing the environment for his present needs. This brings a threat posed on the anthroposphere and on the natural environment alike, interacting with humankind through biological and chemical processes that strain the carrying capacity of the existing equilibrium between humans and the natural environment. Sustainable development as such may be seen as the result of generating a balance between these apparently antagonistic forces: environmental and human equilibrium. One of the main culprits to generate this antagonism is, in large, economic growth. The excessive and asymmetric consumption human being incurs leads to a sink on the environmental sustainability, observed by systemic changes of the landscape. Examples of this strain are rapid land use changes found in urbanization processes, loss of ecosystems and biodiversity, and exploration of available resources leading to scarcity. Regional development tries to attenuate, understand, and generate policies that mitigate the effects of human being on scarcity and therefore rely on a sound understanding of land use and landscape. This keystone species role of human being is nested in the capacity of creating policies on the limits of environmental exploration, as well as informing and conceptualizing our ecological footprint as a species. The notion of territory and sustainable development, at a regional scale, gains in this context a great importance (Fig. 3.4). Regions and local communities can adopt a decisive role on endogenous growth, and local communities may shape social behaviors at a global scale through internal local actors.

In this sense, the role of historical and archaeological landscapes is dual, on one side boosting consumption by using available historical landscapes as a resource and guaranteeing human economic growth while avoiding the negative externalities brought by economic growth. Examples of these include, among others, the refinement of sustainable tourism and exploration of agricultural sectors while fomenting the local creative class. The nurturing of the creative class results in social and cultural awareness that from a spatial perspective is intrinsically linked to the territorial memory. Furthermore, when aided by spatial information systems and ambient information systems, the local values are shared as a knowledge base throughout the world and foster examples of a living consciousness of global sustainable development through regional science. I do assume that we are nowadays witnessing what may be designated as "collapsing landscapes," shared by a postmodernist vision of fragmentation of (1) land use, where the landscapes have been afflicted by such profound changes in the structure of its land use and environment that heritage landscapes and their contribution to economic growth may become lost for future generations, such as (2) territorial identity (i.e., the memory of the land and of the traditional sectors of economic and social activity has irreversibly become lost with the disappearance of historical, archaeological legacy) and (3) memory, where the land use and historical identity share an affective value which society takes for granted, as the process is slow, steadily changing the landscape. This leads to assume that the available spatial information of archaeological sites and natural heritage in a region should be considered as an indicator of sustainable human habitats resulting from a transversal integration of disciplines. This results in two distinct future visions on the interaction of landscape and human heritage from a spatial systemic perspective that may represent pathways to avoid the fragmentation mentioned earlier: (1) coherent landscape, where urban planning plays an important role in maintaining a balance between the urban environment, the regional environment, and the physical structure of the city and its hinterlands. On the other hand, (2) the dominant landscape is applicable to cities registering which are witnessing strong regional conflicts at present. These conflicts are a result of war, policy, and politics and often lead to complete destruction of landscapes of the past.

3.5.1 *The Coherent Landscape*

Within the notion of changing regions, the values of landscapes are measured by the transitions of land use and land cover. A sign of progress in terms of sustainable development for the coherent landscape is the least amount of changes on the ecosystems, maintaining the memory of the region as an intact and predominantly presence in shaping the regions of the future. Within what I define as a coherent landscape, urbanization processes remain fundamentally static over time. The coherent landscape depends on the integration and synergy of ecosystems and biodiversity for human being while maintaining a sense of order. In this sense, urban regions and rural areas are integrated into a homogenous platform, where small cities are stimulated instead of large urban regions. The landscape is thus strictly functional and depends on the local and regional resources and is self-sufficient. The coherent landscape is further explored by the creative class, taking advantage of innovation within small clusters of expertise, that contributes to creating a territorial identity where values are shared, but remained continuous over time. This spatio-social coherence in the landscape is transmitted to urban areas and should provoke fewer changes on land use, as well as assimilation of historical landscapes and archaeological landscapes as a part of the social and economic product the region offers. From a regional modelling approach, this suggests an integration of Shannon's entropy for what can be derived as spatial entropy. The concept of spatial entropy binds (1) an associated measure for the regional spatial extent and (2) a functional reference to a geometric or a geographic coordinate system. As such the derived formulation of entropy for a discrete case, written as functions of probability densities p_i where spatial variability may be written as follows:

$$H = \lim_{\Delta x_i \to 0} - \sum_i p_i \ln\left(\frac{p_i}{\Delta x_i}\right)$$

where Δx_i is the spatial interval size of a given zone i within the region and the study area and p_i is the probability of a given event. Entropy, particularly the spatial manifestation within Shannon's entropy, becomes thus a measure to correlate urban sprawl within the coherent landscape. In fact, entropy represents thus spatial concentration or dispersion over a specific scale which may be assessed at the regional level. This leads to the coherent landscape, where low entropy values thus pertain to regions which the landscape has witnessed less sprawl and the creation of smaller yet more sustainable cities.

3.5.2 *The Dominant Landscape*

The dominant landscape pertains as part of the conflicting structure of rapidly changing regions. The structure of these cities is to fundamentally re-equate the regions based on growth strategies and limit radically the possibilities of urbanization. In this sense downscaling of economic production should minimize throughput and salvage ecosystems. Limiting urbanization in areas where accelerated processes have been witnessed over the last decades at the regional level should be considered. Within the dominant landscape, where urban pressure seems to be eminent, heritage must serve as a containment factor to preserve the authenticity of the current landscape. Heritage endangerment is thus seen as a tool to measure the fragility of the area that informs decision-makers of the limits to growth within these regions. Further, the dominant landscape is characterized by the existence of a dominant species, human being, whose impact at the regional level leads directly to loss of biodiversity and direct impacts on the environment as well as its sustainability. In this sense the dominant landscape strongly links to anthropogenic activity and coping with sustainable development at transborder regions affecting at the regional level the regional impact of its ecological footprint. The issue of coping with the carrying capacity of the landscape must thus not only be assessed within the geomorphology of land use and urban form but, rather, should extend to the impacts of ecological allowance of businesses and regional interactions of institutions and their responsiveness to sustainable development.

3.5.3 The Vertical Landscape

Continuous urban growth and concentration of economic activity is leading to changing regions due to urban development, where economic geography and the importance of local in livability of areas are important aspects of creating a more sustainable future. A place-based approach answers to the proxy of continuous economic growth, and concentration of urbanization with better coordination of transport, land use, and open space planning may lead to sustainable development megacities. The vertical future of the megacities' sustainable landscapes, from a regional science perspective, further enhances the possibilities of economic prosperity while leading to an increased technological efficiency reducing costs in terms of transportation and accessibility to information. The city-dwellers of the vertical landscape must consider the potential of building high-rise urban regions, maximizing the accommodation of infrastructure while avoiding the cost of sprawl.

3.6 Geovisualization: The Role of Mentoring Regional Science in the Future

The different forms the landscape may reflect in the future are largely a result of the possibilities to visualize and calibrate our socioeconomic and environmental variables as well as the availability of data sources today. The spatial component is quite relevant for understanding the changes witnessed at land use level and becomes even further important as to understand the complex interaction between human being and its environment, particularly given the rapid changes in the urban world. Geographic information systems and in particular geovisualization will have an even larger role in future building scenarios and measuring hypothetical outcomes throughout geographical space. This role is boosted by the capacities of computing future scenarios of urban and regional change for sounder decision-making, as well as social awareness of regional social interaction of information with the advent of web 3.0. The changing landscapes will rely heavily on technological advances to promote sustainable behaviors and allow population dynamics to act based on information. At the regional level, the role of geovisualization will be strongly linked to developing sustainable development through identification, use, and dissemination of spatially explicit information. The integration of information for decision-making within the framework of web 3.0, centering around the availability of data and social participation, leads to a more collaborative role of geovisualization and intelligent regions. Concerning regional growth this may lead to a more equilibrated society, where resource use is shared and optimized, supporting a more functional region. In this sense, information becomes the driving force of an autonomous functional society, where resources of habitation and spatial understanding of regions promote a better living quality as a whole. While at present the role of geovisualization is defined as a tool to assist in understanding the geographical dimension of the region, in the future this role will be shared to a collective sense, of jointly changing the region to become more functional for the individual promoted by ambient information systems, where people interact and relate to digital information in the context of their environment. This ubiquitous role of spatial information is accelerated by the recent advances in crowdsourcing especially through crowd-designing, where urban interactions can be used to converge top-down and bottom-up approaches setting up a live knowledge network. This live knowledge network must be linked strongly to the regional decision-making agenda. While geographic information systems alone can only offer a limited number of solutions, it is through the creation of transversal efforts in regional science that the understanding of landscape and its complexity becomes possible. The advances of using spatial information combined to regional sustainable development, and understanding of socioeconomic interactions, can allow creating a new agenda for regional science, where the focus is of transmitting to future generations the changes beginning at present.

The recent emergence of the importance of the multidisciplinary contribution of geodesign, as a support system between planning and design, creates a central role in assessing the landscape. The efficiency of landscape within the structural benefits of smarter options brought by better design solutions can integrate thus better decision support systems that allow optimization at different scales. The integration of GIS components leads to the availability of more efficient tools that may readdress traditional techniques found in cartography through geodesign. Additionally, the component of visualization brought by cartographic representation together with spatial analysis and the existing geocomputational advances bring the potential to redesign regions, thus shedding an unprecedented contribution to regional sciences. The integration of these components creates a framework for regional intelligence, where socioeconomic and demographic data can be integrated within the fusion of geodesign and the plethora of tools found in geographic information systems and fostered by the diversity of socioeconomic and multi-temporal data (Fig. 3.5).

This integrative approach to regional intelligence suggests that the current advances in different typologies of modelling approaches can be efficient contributions to (1) assess, (2) visualize, and (3) intervene in the process of landscape planning particularly for urban regions. The incorporation of large data in GIS allows thus an exploratory prospect within geovisualization, while the assessment itself of these more analytical models benchmarks tools for social awareness as well as a direct contribution to available information through participation. Volunteered geographic information (VGI) is a promising example of this contribution. The information building brought by more robust spatial data at different geographical

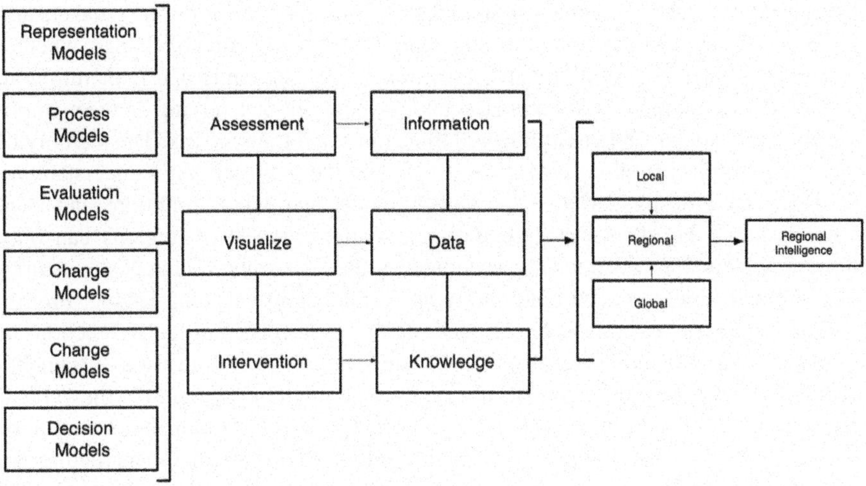

Fig. 3.5 The framework for regional intelligence

scales brings thus the opportunity for social intervention and fosters the opportunity for community engagement leading to local knowledge. The distinct dimensionality that exists over geographical scale, either local, national, or global, makes regional solutions particularly interesting, given the adaptability and the potential at the regional level to consider better integration of larger aggregation of socioeconomic variables also for policy. The modelling capabilities extend thus into the potential of answering complex local as well as global paradigms into the field of regional decision-making (Vaz et al. 2015) and promote the new concept of regional intelligence.

3.7 Conclusions

As much as our impact on Earth has brought irreversible environmental change, our landscapes have in detriment of these choices witnessed a substantial change, most of it affecting our natural and historical heritage. In the context of regional development, economic geography and complex space-time dynamics are factors of continuous change. Monitoring of the transitions of land at the regional level is thus of utmost importance for sounder regions in the future. It is relevant to preserve landscapes by enabling efficient economic growth, without jeopardizing the natural ecosystems and mitigating the impacts on the anthropogenic heritage and archaeological landscapes alike. This paper has shown the possible spatial interpretations of landscape change by means of defining the role at present of geographic information science and systems as tools to allow sounder urban and regional interactions. The proposed three pathways integrating regional development within a spatial landscape preservation framework should serve as a guideline for policy makers and regional planners to note future development of regions under rapid change, whether economic, social, or environmental. Functionality of the landscape is therefore an utmost important trade of the future directions of our regions, relying on the memory of economic, social, and environmental transitions and transactions. The memory of these spatial transactions is possible to archive in the multi-temporal understanding of what I have designated as the collapsing landscape, where traditions and local culture should be always catered. From a spatial perspective, regions can only become sustainable, when what was defined as spatial memory – that is, the identity of place and time and economic traditions – is coherent and long-lasting. From a regional science perspective, we can hence define three epistemological directions of the landscape paradigms within the collapsing landscape leading to a sustainable future: (1) the coherent landscape, (2) the dominant landscape, and (3) the vertical landscape. All of these types of landscapes largely depend on our options taken in the next decades. The usage of geographic information systems, in particular the recent advances in location-based services, crowdsourcing, and ambient information, may act with a leading role in the development of regions and may act as a visual tools for evaluation of landscape change, enhancing our sustainable future.

References

Arsanjani JJ, Tayyebi A, Vaz E (2016) GlobeLand30 as an alternative fine-scale global land cover map: challenges, possibilities, and implications for developing countries. Habitat Int 55:25–31

Bellout A, Vaz E, Penfound E (2020) Rethinking agricultural land use in Algiers: a spatial analysis of the eastern Mitidja plain. Habitat Int 104:102239

Capello R (2001) Urban growth in Italy: economic determinants and socio environmental consequences. Report 10:2001, CERUM

de Noronha Vaz E, Cabral P, Caetano M, Nijkamp P, Painho M (2012) Urban heritage endangerment at the interface of future cities and past heritage: a spatial vulnerability assessment. Habitat Int 36(2):287–294

Esteves TC, Alves FL, Vaz E (2020) Coupling agent-based Modelling with geographic information Systems for Environmental Studies—a Review. In: Regional Intelligence. Springer, Cham, pp 225–249

Fischer MM, Getis A (2010) Introduction. Springer, Berlin/Heidelberg, pp 1–24

Monteiro V, Painho M, Vaz E (2015) Is the heritage really important? A theoretical framework for heritage reputation using citizen sensing. Habitat Int 45:156–162

Onilude O, Vaz E (2020) Data analysis of land use change and urban and rural impacts in Lagos state, Nigeria. Data 5(3):72

Vaz E (2014) Managing urban coastal areas through landscape metrics: an assessment of Mumbai's mangrove system. Ocean & coastal management 98:27–37

Vaz E (2016) The future of landscapes and habitats: the regional science contribution to the understanding of geographical space. Habitat Int 51:70–78

Vaz E (2020) Archaeological sites in small towns—a sustainability assessment of Northumberland County. Sustainability 12(5):2018

Vaz E, Nijkamp P (2015) Gravitational forces in the spatial impacts of urban sprawl: an investigation of the region of Veneto, Italy. Habitat Int 45:99–105

Vaz E, Cusimano M, Hernandez T (2015) Land use perception of self-reported health: exploratory analysis of anthropogenic land use phenotypes. Land Use Policy 46:232–240

Vaz E, Taubenböck H, Kotha M, Arsanjani JJ (2017) Urban change in Goa, India. Habitat Int 68:24–29

Vaz E, Shaker RR, Cusimano MD, Loures L, Arsanjani JJ (2020) Does land use and landscape contribute to self-harm? A sustainability cities framework. Data 5(1):9

Chapter 4
Analytical Tools from a Socioeconomic Point of View

Abstract This chapter enhances the previous discussion to point out some of the most important arguments to understand Southern Europe's challenging reality better. Thus in this context, we proceed by presenting a sequence of analytical methods and tools that have been used to detect some of the significant features able to readdress pushing and pulling forces, those that could be considered the most efficient for change in a dynamic process of the region. This chapter launches the framework for dynamic analyses by proposing a conceptual context to describe and compare industrial models and production systems across southern European regions, thereby emphasizing the need to use analytical methodologies before building strategic decisions for future developments.

Keywords Analytical methods · Regional growth · Dynamic growth patterns · Southern Europe · Industrial models · Production systems

4.1 Framework for Dynamic Analyses

The importance of creating a framework for dynamics analytics is essential and should always be in line with a thorough evaluation of conditions as well as strategic policy design. This chapter draws to southern Europe as an empirical example focusing on local identities and regional cooperation as strengths that should be pursued for southern Europe's continuous economic and environmental prosperity.

Since the middle ages, southern European history has brought many experiences of decentralization and regional survival, particularly in Spain and Italy, providing institutions with driving contexts for decentralized analytics. Whenever centralism has been imposed for political reasons, the expression of territory's identity emerged proceeding from local initiatives of different kinds and building up new spaces or new solidarity concepts, stressing for new authorities in defense of the application of decentralization policies and the management of new rules. Examples of this may be found at a political level in the Basque Country and a socioeconomic level in Cataluña, in Spain.

E. Vaz, T. de Noronha, *Sustainable Development in Southern Europe*,
https://doi.org/10.1007/978-3-662-62177-6_4

From an economic standpoint, the Italian Marshallian industrial district was the starting point for a comprehensive analysis of the local development phenomena based on how technological change could enhance regional identities and prosperity. The scientific community had accepted the interaction between technology and territorial dynamics either by the evolutionist view of learning (Dosi 1988) or by endogenous development (Barquero 2007) and the approach to the innovating environments (Maillat 1996).

4.1.1 The Concept of Local

Any geographical space is subdued, within a given period to dynamic processes resulting from the changing relationships among agents that define social and economic behavioral patterns. Williamson (1985) had defined such relationships as transactions, arguing that they could be identified as external to the markets or, instead, internal, hierarchical, and, sometimes, even cooperative.

The way these transactions occur may design the development path of the geographical space where stockholders interact. These are layered by a sociocultural structure that encourages the existing knowledge capacity that we define as territory. Recent globalization movements have determined many unexpected new market rules and trends. Most of them defined for territories a set of new, almost imposing external relationships to be adopted. Today, no geographical space can exclude the importance of external relationships in the design of their development processes, whether geared to the market or internal to their hierarchical structures. This assumes that cooperation remains the adequate relational choice for geographical dynamics and that no local level can be prosperous without the commitment to further connect to adjacent regions. Spatiotemporal understanding has become a key driver for territorial dynamics, resulting from stakeholders supporting structural change while acquiring the capacity of knowledge dissemination and regional interaction. With this new synergy across regions, it becomes noteworthy to assess how the capacity of knowledge dissemination may be introduced within a finite regional scale.

The conceptual idea of knowledge dissemination furthers the question of how structural change happens in the territory. How does the territory consolidate its absorptive capacity to be able to integrate new organizational or technical procedures favorable to innovativeness? Possible replies to these questions require a clear understanding that territories do not incorporate a spatial dimension uniquely, but rather include parameters that evolved within historically. Institutions function as regulators of individuals as well as collective practices. In such conditions, the significant resources become relational, as explained by Sierra (1997), and proximity among the stockholders converts in the major asset of territories (Breschi and Malerba 1997).

The issue of adjacency as a knowledge dissemination factor of integration of geographical and technological proximity deserves further attention. The notion of geographical proximity suggests increases in the relation between economic activities, heightening the idea of organizational and institutional proximity. Technological proximity, on the other hand, pertains to the necessity of interaction among agents, primarily when resulting from knowledge dissemination and learning.

In a context of increased market competition and accelerated innovation, firms and regions search for competitive factors rather than price advantages such as the capacity to systematically acquire new skills. A strong capability to learn and transform organizational competencies is thus highly advantageous. The "learning region" favors pre-conditions for learning in the forms of common regional culturally based rules or cooperation, often with collective agents. Its recently upgraded concept of "smart regions" promotes the awareness of the existing skills and territorial identity to focus on adequate learning strategies.

It becomes relevant to position how smart regions manage to optimize their strategies within the collective learning experience and considering the development of new know-how that may be replicated within the territory. Maillat (1998) points out to the concept of "retournement" as an autonomous mechanism, in which each territory changes its hierarchical geographical relations. If there is a hierarchical system of territories, the development of such territories depends on their contiguity to each other. This integrates well the idea of Tobler's first law of geography where "everything is related to everything else, but near things are more related than distant things" (Tobler 1970) as applied to the conceptual framework of knowledge integration and learning. Hence specific actions occurring within nearby peripheries advance the issues related to territorial development to a central process in the configuration of the global industrial system.

Recent advances in the fields of spatially-explicit network theories (Fischer 2006) point towards the significant change in the interaction of innovation processes and the impact on knowledge spillover effects as pillars in the creation of networks. Striking evidence comes from research carried out by universities, promoting regional knowledge production. This is particularly the case in high-technology industries when the occurrence of spillover effects adds value at the corporate and organizational levels (Fischer and Varga 2003). What remains uncertain, however, is the stability of such spillovers when integrated with non-high-technology industries. It is our opinion that these spillover effects occur not only due to their intrinsic strength but also due to the existing collective learning, determined by (1) levels of local creativity, (2) capability to adapt behaviors, and (3) processes of reduction of dynamic uncertainty and resilience.

4.1.2 The Concept of Local/Regional/Territorial Production Systems

As previously discussed, Storper and Harrison (1991), as well as some other researchers, alternatively proposed the notion of "production systems" without interfering with the significant results obtained by the previously described investigations. For Storper, discussing the territorial dimension requires prior identification of the types of "input-output systems" and "functional cores" of the economic activity. Scott and Storper (1990) (in Storper and Harrison 1991) clarified that similarly to the production units, the diverse input-output systems create external economies of scale or scope if units of production are fragmented and concentrated on a social division of labor processes. Due to this, the systematic and increasing segmentation of productive forms induces a consolidation of interrelationships among the various input-output systems, intensifying the need for production flexibility. Hence, spatial proximity does not only reduce geographical distance but also facilitate information exchange, regulating the level of efficiency of local systems by their response capability to external environments.

These organizational models bring their advantages, costs, and risks. The main advantage is the "network surplus" of scale economies in research and development, commercialization and production, know-how complementarities, as well as strategic synergies among firms. However, one of the risks regards the occurrence of opportunistic behaviors of some partners, who may benefit from the cooperation alone. Some definite limitations are related to a high level of organizational expertise that the model requires, making them prone to higher costs.

The networks of industries can be characterized by the existence of a variety of firms linked by a high level of technical compatibilities as well as significant integration due to the externalities provided (Camagni 1991a, b, 1995a, b) such as aggregate cost functions with increasing incomes and net benefits from belonging to the network. Due to a concern in diversity, it has not been easy to adapt the reality of the productive regional bases to this theoretical conceptualization of territories. The region is seldom homogeneous, and, in most cases, different subsystems coexist inside the same region. Nevertheless, the exercise of coexistence among different subsystems and the regional learning capacity (Malerba 1992) may generate complementarities providing articulation within the context of a strategic framework. We may advance that as Fordist solutions lose strength, local specificities may take advantage when the strategic choices are sufficient to enhance sustained productive systems. Such views evolved in an entirely new theoretical framing for strategic regional development bases upon smart cities as drivers of change across the gapping regions (Caragliu et al. 2011).

4.1.3 The Concept of Industrial Models and Production Systems

The concept of industrial models directly linked to the "regulationist approach" from Boyer and Durand (1998) shaped a novel contribution to the debate about the spatial distribution of production (Lung 1995). It introduced uncertainty and stability as a new basis in which relations could be established under different confidence levels as well as in the different forms of labor relations. Based on two main features, the flexible specialization and the flexible accumulation had, later, a significant contribution to enhance a meso-economic approach as a new sphere framing the industrial models and the geographical context of their productive systems (Noronha Vaz 2011).

Other vectors have in the meantime brought a growing amount of attention in these analytics: (1) The model could not be defined independently from its specific spatial and historical context, (2) in a transition period different industrial models coexist independently of the dominance of one of them, and (3) the models have evolutionary tracks with territorial nuances. Finally, the Fordist model can be associated with the labor division and the post-Fordist model to territorial recomposition.

The "regulationist" arguments challenge the Fordist model for the loss of territorial identities and suggest that new economic status may emerge for "production systems" shaped in the post-Fordist era. Following more flexible organizations that are adapted to the market, and given the development of networks of firms, more suitable settings will combine the local and the global contexts. To better understand the market's expectations, firms relocate, at least one part of its outset and procedures, sending that regime in which supplies are identical for all geographical territories.

4.2 The Need for Applied Methods to Build Strategic Decisions of Development

The previous section focused on the desegregated comparative analyses of data and their contribution to better understand the reality of southern Europe. In this section, a sequence of analytical methods enables us to understand the significant characteristics and changes in European lagging areas.

Western European economies have since the industrial revolution increased on average by 2–2.5% a year (European Commission 1997, 1999), until the early 2000s. However, remarkable disparities between countries and, most notably, between regions can be detected. National statistics on GDP, the standard measure for economic performance, show that there is a gap between the GDP in Spain, Greece, and Portugal compared to the GDP of other European countries. This gap has an origin in the unbalances of certain specific European lagging regions, of which many are in southern Europe, threatening to spread all over Europe as the earlier increasing disparities between some European regions are shrinking the growth capacity of the others.

Identification of territorial systems across Europe uses the NUTS classification (Nomenclature of Territorial Units for Statistics). This classification system was conceived by EUROSTAT and extensively used in Community legislation since 1988. The proposed sample observes several territorial systems, as either NUTS II or

Table 4.1 Territorial systems by NUTS level, terminology, and size

		NUTS level	Terminology	Size (km^2)
1	Alentejo Central (Pt)	NUTS III	Region	7.228
2	Aude (F)	NUTS III	Departments	6.139
3	Northern Border (E)	NUTS III	Region	12.341
4	Cremona (I)	NUTS III	Province	1.770
5	Devon and Cornwall (GB)	NUTS II	Group of counties	10.262
6	Gard (F)	NUTS III	Department	5.853
7	Hainaut (B)	NUTS II	Province	3.785
8	Hereford and Worcester (GB)	NUTS II	Group of counties	3.923
9	Kujawsko-Pomorskie (P)	n.a.	Province	20.099
10	Piacenza (I)	NUTS III	Province	2.589
11	Oeste (Pt)	NUTS III	Region	2.512
12	South West (E)	NUTS III	Region	12.306
13	West Flanders (B)	NUTS II	Province	3.134

Source: National statistics of different countries

NUTS III regions. All of these are selected among different European countries, with varying size from 1.770 km, in Cremona, Italy, up to 20.099 km square in Kujawsko-Pomorskie, Poland, as shown in Table 4.1.

Firstly, two indicators of economic performance are assessed: The first indicator, gross regional product per capita, measures the total output in a specific area, including services. The second indicator is employment by sector: While, in rural areas, agriculture incorporates employment, more urbanized areas have an essential contribution of employment for the service sector. Furthermore, agricultural activities may be providers of ingredients for local food manufacturing enterprises. There is evidence suggesting that small and medium enterprises in the food sector are usually located in rural areas where they have developed by processing products of local agriculture. For the studied regions, data on GDP and employment is presented in Table 4.2.

A more detailed observation of the data presented shows that the territorial system with a higher GRP per capita was West Flanders, a region that fits into the top 25% of the wealthiest regions in the EU, based upon a significant amount (over 5%) of employment in agriculture and an unemployment rate below the national average. Contrarily, Hainaut showed a high GRP per capita, but the economic performance of the territory was low compared to other Belgium provinces. In this case, services played a significant role in employment in Hainaut; however, the unemployment rate was about 13%, the highest in Belgium; the Languedoc-Roussillon region displayed values below those of the European average. In both territorial systems, agriculture and food processing are an essential source of employment, although Languedoc-Roussillon was still lagging when compared with other regions in France.

Piacenza and Cremona show similarities to France. Employment in the primary sector exceeded 7% of total employment compared to Piacenza's industrial sector, exploring activities within the food sector such as pasta, cheese, salami, and the

Table 4.2 Indicators of regional economic performance in different regions across Europe

Indicators	GRP		Employment by sector		
Region	GRP (1.000 euros)	GRP growth 1995–2000	Agriculture (% of total)	Industry (% of total)	Services (% of total)
Belgium			2.7	27.5	69.8
Région Wallonne			2.9	24.5	72.6
Hainaut	16.0	+0.16	2.7	26.9	70.5
Flanders			2.9	30.6	66.5
West Flanders	21.9	+0.18	5.1	33.1	61.8
France			4.6	26.6	68.7
Languedoc-Roussillon			7.8	19.4	72.8
Aude	n.a.	+0.03	8.8	27.2	64.0
Gard	n.a.	+0.03	7.0	38.5	54.5
Ireland			10.9	28.5	60.4
Northwest Border	13.4	+0.23	14.3	35.9	49.7
South West	19.0	+0.34	12.2	29.9	57.8
Italy			6.5	31.7	61.8
Emilia Romagna			6.3	34.6	59.1
Piacenza	18.7	+0.16	7.2	35.5	57.2
Lombardia			2.7	40.6	56.6
Cremona	16.6	+0.13	7.6	40.5	51.9
Poland			–	–	–
Kujawsko-Pomorskie	3.0	+0.07	26.4	31.0	42.6
Portugal			13.3	31.0	55.7
Alentejo Central	7.8	+0.27	14.2	24.5	61.2
Lisboa e Vale do Tejo			3.8	25.2	71.0
Oeste	8.0	+0.18	8.7	38.9	50.0
United Kingdom			1.9	26.8	71.1
South West			3.2	25.4	71.2
Cornwall, Devon	13.5	+0.00	5.0	24.1	70.7
West Midlands			1.9	34.3	63.7
Hereford and Worcester	16.1	+0.08	2.8	29.8	67.2

Sources: Data collected from regional and national statistics, European Commission

canning industry. This is comparable to Cremona, where the traditional construction of farm machinery and musical instruments has specialized endogenous growth patterns of territorial relevance. Both Emilia Romagna and Lombardia have low unemployment rates when compared to other EU regions.

When compared to the other territorial systems, Kujawsko-Pomorskie, in Poland, showed the lowest GRP (5% of total GDP), as well as a low growth of its GDP. In this case, about 25% of the population worked, but an unemployment rate exceeded 13%. Two territories illustrate the Portuguese case with low GRP but register significant growth for Alentejo Central and Lisboa e Vale do Tejo. The service sector is critical in these cases, and Lisboa e Vale do Tejo region also has a vital agro-food industry. Alentejo holds a high proportion of employment in agriculture and a lack of tradition in industrial processes, a case that needs a restructuring of the marketing systems and within the produced food products. Unemployment rates of both regions befall in the same category.

In the United Kingdom, our case studies illustrate moderate GRP per capita. The South West region is one of the UK's best-performing and most vastly growing regions. Still, considerable subregional disparities are patent. The north and east of this territorial system are successful; however, added value to wealth and ability to generate new startups are somewhat divergent from the isolated regions such as Dorset, Devon, and Cornwall. For example, the GRP of Cornwall is the second lowest in England. In such cases, also the primary sector sustains 5% of employment as the traditional industries have suffered substantial decline. The primary industry in Cornwall was formerly tin mining, but the last mine closed with the increase of low prices from global markets. Other activities such as agriculture, fishing, and china clay extraction have also reduced. Nevertheless, traditional industries persist as essential pillars of these regional economies. In Hereford and Worcester, GRP is higher than in Cornwall and Devon, but its growth rate is quite low. The largest employment sector in Hereford is food and non-food manufacturing. Almost two decades ago, the employment structure in Worcestershire was primarily made up of three sectors that account for 75% of all employees in Worcestershire: manufacturing; public administration, health, education, and distribution; and hotels and catering (Noronha Vaz et al. 2001).

As to characterize the different territorial systems, a simple methodology is offered by assembling all the 13 observed regions based on a set of 17 indicators out of a composition of 20 preliminary variables. In opposition to the simplistic criteria related to the exclusive use of the level of the GDP, and since we have abundant and reliable data, multivariate statistical analysis can be applied. This method shows the benefit of admitting as many variables as necessary to theoretically better approach the significant determinants of regional growth in each one of the different territorial systems.

Numerous considerations may emerge from the variable selection, and few restrictions apply:

1. The model must permit the integration of a number of variables higher than the number of the observed regions.
2. The variables should be obtained for each region and observed for the same period. This may present some obstacles, as there often lacks a formal standardization process between countries.
3. The observations timestamps are not coincident for all countries or regions.

Further to this, the following step submits the date to cluster analysis, followed by a discriminant and correlation statistical inference, which generates a different set of

Table 4.3 Variables of local development

Indicator	Description
IRPpc	Internal regional product per capita
GAV/ Employment	Productivity
GAV 1/Employ-ment 1	Productivity in primary sector
GAV 2/Employ-ment 2	Productivity in secondary sector
GAV 3/Employ-ment 3	Productivity in tertiary sector
LFQsec/active pop.	Labor force qualification (number of students in secondary level) in proportion of active population
LFQhig/active pop.	Labor force qualification (number of students in higher level) in proportion of active population
Trans-com	Investments in transports/communications in relation to surface
% empl 1	Percentage of employment in primary sector
% empl 2	Percentage of employment in secondary sector
% empl 3	Percentage of employment in tertiary sector
New 1	N° of constituted enterprises in relation to total, primary sector
New 2	N° of constituted enterprises in relation to total, secondary sector
New 3	N° of constituted enterprises in relation to total, tertiary sector
Exp food	Proportion of household expenses in food
Exp drinks	Proportion of household expenses in drinks
Exp catering	Proportion of household expenses in catering

Source: Noronha Vaz et al. (2001)

regions, composed by elements the most homogeneous as possible known as territorial systems.

The possibility of observing a minimum of two timestamps of these analytics permits to elaborate a regional dynamic simulation that comprehends regional dynamics of change and possible stages of economic prosperity or recession.

Secondary data may also be used to classify the clusters of the observed regions, and when added to the statistical inference of correlation analysis, it places the most significant variables in the clustering process. Theoretically, there remains a multitude of variables that describe the environmental context of growth within the region, those that can better reflect the local capacity of industrial growth as well as the financial support structure to regional development – the variables of local development (VLD), as shown in Table 4.3.

As an adequate method, cluster analysis has been chosen to classify the regions, clustering these in specific sets composed by elements as homogeneously as possible – territorial systems. As shown in Table 4.4, the analyses were made in 1994 and repeated in 1997, with very different results: In 3 years the determinants of regional growth changed, proving that territorial systems have an unstable nature and respond fast to external or internal impulses.

Table 4.4 Classification of the territorial systems

Cluster	Regions (t)	Regions (t+4)
1	Alentejo Central; Kujawsko-Pomorskie; Oeste; South West; Hereford and Worcester; Border; Devon and Cornwall; Gard; West Flandria; Aude; Cremona	Alentejo Central; Kujawsko-Pomorskie; Oeste; Devon and Cornwall; Border; Hereford and Worcester; South West; Gard; Hainaut; Aude; West Flandria
2	Hainaut	Cremona
3	Piacenza	Piacenza

Source: Noronha Vaz et al. (2001)

Discriminant analysis indicates the average value for each variable in each group that the regions belonging to cluster 1 are more reactive to the following variables: productivity levels, labor force qualification (secondary level), the proportion of employment in primary sector, proportion of new enterprises in secondary sector, and the proportion of household expenses in food. Cluster 2 is strikingly different from the drivers in the regions of cluster 1: productivity levels in primary and secondary sectors, labor force qualification (high level), the proportion of employment in the tertiary sector, proportion of new enterprises in primary and tertiary sectors, and proportion of household expenses in alcohol. As for cluster 3, the following variables express to be the most significant: levels of development in terms of IRP per capita, productivity levels in the tertiary sector, transports and communications, proportion of employment in the secondary sector, and proportion of household expenses in catering products.

This very same discriminant analysis has been used 4 years later, and, although the clustering groups did not suffer significant changes, the drivers for growth changed substantially. Regions belonging to cluster 1 showed some new drivers for change, such as productivity levels for primary and secondary sectors, the proportion of employment in primary and tertiary sectors, the proportion of new enterprises in secondary and tertiary sectors, and the proportion of household expenses in food and drinks. Changes in regions belonging to cluster 2 depended mostly on the proportion of employment in the secondary sector. For the regions grouped in cluster 3 IRP per capita, levels of productivity (in general and in the tertiary sector), labor force qualification (secondary and high level), transports, and communications, the proportion of new companies in the primary sector became important.

Throughout the 4-year period, determinants of change in cluster 1 are dependent on a multitude of drivers such as productivity levels (in general as well as the tertiary sector) growth of expenses in transports and communications, proportion of employment in secondary sector, proportion of new enterprises in tertiary sector, and proportion of household expenses in alcoholic beverages and catering products. The key variables of cluster 2 correspond to the qualification of the labor force (highly qualified personnel), the proportion of employment in the primary sector, the proportion of new companies in the secondary sector (showing a mild decrease in the growth rate), and the portion of household expenses in food.

Finally, cluster 3 shows an increasing rate in IRP per capita, levels of productivity for the primary and secondary sectors, labor force qualification (secondary level), the proportion of employment in the tertiary sector, and proportion of new enterprises in the primary sector as main determinants.

The use of correlation analysis allowed to identify directly as well as inversely correlated variables to gain a clearer perception of the level of closed relationships among some selected variables such as the internal regional product, global productivity, highly qualified labor force, investment in the transport sector, employment by activity sector, and creation of new companies, exemplified as follows:

1. *Internal regional product* was directly correlated with productivity in industry, agriculture, and services as well as global productivity, for the first time period. However, during the preceding 4 years, significant changes happened, and labor force qualification and employment in services became highly correlated as well. Inverse constant correlations have been detected between employment and food exports. Additionally, intriguing is the fact that the creation of new industry in the service sector showed a high correlation to agriculture production.
2. *Highly qualified labor force* showed to be a variable of interest, as the number of directly correlated variables is decreasing over time. During the first timestamp variables such as labor force qualification, the export of beverages, the employment in services, the formation of new companies, and the internal product were significantly correlated. During the second timestamp, however, these direct correlations significantly decreased, only showing a moderate correlation in the exportation of beverages, labor force qualification, and internal product.
3. *Investments in transportation* are not inversely correlated with any of the other considered variables, also considering the temporal dynamics, suggesting that increases in such investments can only generate positive impacts on any other variable.
4. Concerning *employment* in agriculture, industry, and services, the results indicate that direct correlations occur in the first and secondary sectors due to food exportation and the creation of new companies, quite differently from the tertiary sector. It has also been noticed that employment is linked to skills and labor force qualification.

   In the last timestamp, it has been shown that investment in transports substituted the importance of labor force qualification. Inverse correlations, however, demonstrate that increases in employment in industry and services and global productivity or agricultural productivity do indeed lead to less employment in the agricultural sector.
5. *The creation of new companies* was the last group of variables assessed. Indeed, groups of different factors seem to determine the nature of new companies in each of the three sectors of economic activity. For the primary industry, agriculture, and for the first timestamp, productivity, investments in transportation, creation of new companies for the service sector, and highly qualified personnel are extremely correlated variables for the successful establishment of companies. In the case of industry, however, the positive correlations report to labor force

qualification secondary, which is quite interesting to observe. Finally, concerning the service sector in general, productivity (either global or by industry sector) is the most correlated variable.

The results for the 13 observed regions that vary over time recognize that the dynamics of growth are continuous drivers, also in the rural and peripheral regions of southern Europe. In southern European regions, however, drivers of change are not uniform nor continuous but highly irregular. These are quite notable characteristics that should be considered in the decision-making process by policy makers. It is of great importance to design tailor-made support systems that pertain instruments that efficiently measure development, having into consideration the irregularity for these regions and offer thus customizable solutions.

References

Barquero AV (2007) Desarrollo endógeno. Teorías y políticas de desarrollo territorial/Endogenous development. Theories and policies of territorial development. Investigaciones Regionales 11:183–210

Boyer R, Durand JP (1998) L'après Fordisme, La Découvert and Syros Alternatives Economiques, Paris

Breschi S, Malerba F (1997) Sectoral innovation systems: technological regimes, Schumpetrian dynamics, and spatial boundaries. In: Edquist C (ed) Systems of innovation, Pinter, pp 130–156

Camagni R (1991a) Innovation networks: spatial perspectives. Belhaven Press, London

Camagni R (1991b) Introduction: from the local milieu to innovation through corporation networks. In: Camagni R (ed) Innovation networks: spatial perspectives. Belhaven Press, London/New York, pp 1–9

Camagni R (1995a) Espace et temps dans le concept de milieu innovateur. In: Ralllet A, Torre A (eds) Économie Industrielle et Économie Spatiale. Economica, Paris, pp 193–210

Camagni R (1995b) Global network and local milieu: towards a theory of economic space. In: Conti S, Malecki E, Oinas P (eds) The industrial enterprise and its environment: spatial perspectives. Avebury, England, pp 195–213

Caragliu A, Del Bo C, Nijkamp P (2011) Smart cities in Europe. J Urban Technol 18(2):65–82

Dosi G (1988) Sources, procedures and micro-economic effects of innovation. J Econ Lit XXVI (September):1120–1171

European Commission (1997) The globalising learning economy: implications for innovation policy, 175p

European Commission (1999) Sixth periodic report on the social and economic situation and development of the regions of the European Union. Brussels

Fischer M (2006) Innovation, networks, and knowledge spillovers: selected essays. Springer, New York

Fischer M, Varga A (2003) Spatial knowledge spillovers and university research: evidence from Austria. Ann Reg Sci 37(2):303–322. Springer, New York

Lung Y (1995) Modèles Industriels et Géographie de la Production. In: Ralllet A, Torre A (eds) Économie Industrielle et Économie Spatiale. Economica, Paris, pp 85–110

Maillat D (1996) Systémes territoriaux de production et millieux innovateurs. In: OCDE, Réseaux d'entreprises et développement local, pp 75–90

Maillat D (1998) Innovative milieu and new generations of regional policies. Entrep Reg Dev 10(1):1998

Malerba F (1992) Learning by firms and incremental technical change. Econ J 102:845–859

Noronha Vaz T (2011) The design of industrial models: addressing cooperative behaviours, innovation and public policy. In: Desai S, Nijkamp P, Stough R (eds) New directions in regional economic development: the role of entrepreneurship theory and methods, practice and policy. Edward Elgar Publishing, Cheltenham

Noronha Vaz MT, Cesário M, Avermaete T (2001) Territorial systems in the rural areas of the European Union. Confidential Report, Contract n° HPSE-CT-1999-00024

Sierra C (1997) Proximité(s), Interactions Technologiques et Territoriales: une revue. Revue d'Économie Industrielle 82:7–38

Storper M, Harrison B (1991) "Flexibility, hierarchy and regional development: the changing structure of industrial production systems and their forms of governance in the 1990s". Res Policy, n°20, pp 407–422

Tobler WR (1970) A computer movie simulating urban growth in the Detroit region. Econ Geogr 46(sup1):234–240

Williamson OE (1985) The economic institutions of capitalism: firms, markets, and relational contracting. Free Press, New York

Chapter 5
Behavioral Patterns of Innovation in Lagging Regions of Southern Europe

Abstract In this chapter, the introduction of time as a factor of territorial dynamics allows transforming simple relations among stakeholders into lasting interactions able to create new requirements, new procedures, or even new routines. Mapping behavioral patterns are not only a title but a necessity that allows us to perceive how stakeholders in southern Europe interact to innovate. They are encouraged by the common objectives of the productive branch to which they belong and permit, thereby creating scope advantages. So far, little has been made in such a context. In our opinion, creating cluster advantages in southern Europe have not been a strategic choice, and we wish to call the attention for this possibility which, if applied in an integrated context of public policy and practice, could promote prosperity across this enlarged region.

Keywords Dynamic growth models · Clusters · Public policy · Southern Europe · Regional sustainability

5.1 Introduction

The recent interest in identifying patterns of business activities, and related regional attractiveness, results from an accelerating intensity of networking within the context of globalization. In itself, the globalization process imposes reduced profit margins. Due to increasing competition, companies no longer have the capacity to jeopardize resources. The need for cost optimization in a setting of global geography emphasizes the importance of transportation efficiency, but also led to a reduction of geographical barriers. This affects smaller industries and specialized local production for better and worse: Competitive business advance and weaker firms do not survive within the business landscape. This situation has encouraged different discussions on how peripheral areas may attract dynamic companies promoting local or regional development. The solutions for their growth depend on (i) transportation and communication networks, (ii) a qualified workforce, and (ii) an innovative and dynamic environment as the best drivers for innovative firms to guarantee their market shares. It remains to be shown if southern Europe

E. Vaz, T. de Noronha, *Sustainable Development in Southern Europe*,
https://doi.org/10.1007/978-3-662-62177-6_5

has an intrinsic capacity for subsistence and prosperity, considering the behavioral patterns and performances of its corporations. This chapter examines the multiple methods able to identify both the regional ecosystem's attractiveness and the behavioral patterns of business.

One should emphasize the general context for the success of corporations, especially for the small businesses in southern Europe. A closer inspection of Portugal's textile production offers an analysis of a vulnerable industrial sector, characterized by high geographical concentration. The sector remains strongly linked to the networking systems and part of agglomerations of capital and labor. These conditions facilitated a steady involvement in international subcontracts for production across the world (Scott 2006). With the emergence of ICTs, the propensity of industries such as textiles (or the agro-food), to succeed in international markets, can quickly increase and, simultaneously, regional promotion can occur (Cesário and Vaz 2011).

On the other hand, consequent positive effects are not necessarily expectable due to the persisting advantages of high-cost regions. These regions rely on a combination of inputs resulting from new technological trajectories with path dependency advantages. In the long run, they may hold some territorial inputs as we can note that the process of globalization leads to bridge tacit to codified knowledge. As described by Maskell and Malmberg (2006), tacit knowledge, provided by local skills, can gradually become codified and approach open markets. Such transformations have often occurred when considering traditional quality products. So, as soon as new technologies and new organizational designs become available globally, firms in low-cost areas become more competitive. Additionally, every time a local input joins the productive process, the firms may react in two distinct ways. Either these firms reduce cost by the relocation of manufacturing production activities leading to unavoidable loss of employment at the regional level or by creating more knowledge, forcing the creation of new local inputs utilizing new and more tacit knowledge inputs. For regions of southern Europe, the best response strategies to consolidate regional resilience should count on the creation of concrete and specialized regional resources. Local actors are being permanently aware and able to manage ecosystems so that companies could become more resilient to a predatory global context as referred by Vaz et al. (2018). In many of such regions in southern Europe, regional competitiveness is achieved by the global flow of ideas, capital, goods, and labor; the role of proximity in the creation of economically useful knowledge appears to be even more important than elsewhere (Storper and Scott 2009). The role of universities and research institutes applying scientific results is probably the best of all tools for consolidating, growing, and promoting the ecosystems of peripheral regions across southern Europe.

At present, distance is still important in many regions, simply because we must assume that the diffusion of information and codified knowledge will occur much faster than the diffusion of understanding and subsequent implementation of capacities. Also, some types of knowledge travel more easily than others. Therefore, if the agglomeration factor is important for business activities because it facilitates transactions and opportunities to match needs and capabilities, also firms' connections may lead to denser local skills and facilitate arrival of localized specific relational assets to promote learning and innovation.

The advantages of location proximity go beyond transactional aspects and include other positive externalities such as knowledge spillovers, rules, or customs that empower firms to coordinate under uncertainty. Technological change must be path dependent, simply because it includes interdependencies between choices made over time. Indeed, the spatial dimension and input-output relations taking place may also affect the organizations to follow technological processes. The established interdependencies that are untraded and include labor markets, conventions, common languages, and rules become the main reason for economic resilience, representing the aptitude of regions to react to the challenges of globalization and modernity.

The earlier discussion on path dependency, as highlighted by Tsipouri (2017), becomes even more relevant when considering the specificities of small businesses in southern Europe. Differently from big enterprises, small and medium enterprises (SMEs) interact intensely with the territory in which they locate, as a sign of their embeddedness. The unusually strong links they develop with the external environment also reduce risks of uncertainty. In general, SMEs do not only locate near the residence of their business owners but also establish geographical and social networks as their primary sources of assets and information.

5.2 New Shapes of Business Organization

During the 1970s, market stagnation and the rise of consumer awareness created unprecedented opportunities for small markets. Niche markets, whose product competitiveness was increasingly measured by its quality rather than by its price, significantly grew (Becattini and Rullani 1995). Traditional companies following a Taylor production model gave room to new corporations that were characterized by innovation, an open management system, and an emphasis on core organizational values. Such companies also held flexible production systems, efficient processes, just-in-time productions, synchronic product development, and total quality management. These were not necessarily small firms. Some were fragmented companies capable to better cope with the implementation of just-in-time production. Of these, the most dynamic corporations had the possibility of producing a wide variety of products in short periods, increasing their cost-efficiency and reducing their stock requisites. Such strategies centered around cost reduction, employing organizational innovation, and captivating markets that did not necessarily invest in quality in their production circuits.

A decade later, quality assurance became a corporate concern. Processual integration and organizational innovation were added in most of the food business and other similar sectors, as well as the increasing use of ICTs, allowing, in return, an objective assessment of the quality of production and significantly changing entrepreneurial attitudes and philosophy.

These new improvements, occurring at the organizational level, had consequences for industrial location across the world: units of large-scale, vertical production that often lead to regions specializing in each sector of high industrial density progressively generated more segmented production lines. These were integrated by sectors that demonstrated increasing geographical flexibility. A more significant geographical redistribution of work connected regions at a global level through management links hiding a critical impact. Earlier market-driven industrial regions quickly vanished, and many big companies and a rising number of small and medium enterprises changed the architecture of their spatial topology from a single geographical location to multiple geographic locations in several regions across countries.

5.3 Implications at the Organizational Level of Industrial Space

Both flexible production and just-in-time production systems led firms to get involved in much more complex business networks, imposing the importance of competitive efficiency for smaller businesses. Also, companies' cost structure (Eliasson et al. 1990) altered, giving room to technological advances in the direction of flexible production. While the internationalization of products causes growing costs related to external marketing and logistics, improved transportation conditions associated with just-in-time production systems generated decreases in transportation and stock management costs.

Considering the solidification of this general context, it becomes essential to identify the local/territorial capacity to attract or embed firms (Noronha Vaz et al. 2003). Next, we present an empirical study taking place across an enlarged set of southern European regions of Portugal, Greece, Spain, and southern France. The supplied analyses identify the main factors of development and select those that best attract innovative companies.

This experiment used a combination of classical variables to categorize different levels of attractiveness of regions as essential characteristics for these regions: (i) quantity and quality of access routes, (ii) workforce and workforce flexibility, (iii) number of personnel, (iv) R&D expenditure and jobs, (v) the distribution of gross added value (GAV), and (vi) gross domestic product (GDP) per capita. The dataset included also less common variables, related to transportation and specific characteristics of labor force: (i) railway network density, (ii) highway network density, (iii) the ratio of active population over total population, (iv) population density, (v) non-active population ratios, and (vi) number of students attending school at each stage of the education process concerning the total population.

The population was categorized as follows: young individuals under 25 years, adults between 25 and 64 years, and seniors over 65 years. Although most regions of developed countries are gradually aging, this process is more advanced in some regions than in others. Also, the workforce and the ratio between part-time workers and actives to indicative of the degree of workforce flexibility was assessed. Finally, research and innovation played a significant role as variables such as the effort provided by R&D, R&D expenses in GDP percentage terms for each region, as well as R&D's job percentage of actives have also been considered. Further, a distinction was made between the genesis of the job and R&D expenses, i.e., business sector, state, and higher state education. Lastly, GAV distribution and per capita GDP were used as an indication of the degree of current regional development. This assessment brought interesting analytical conclusions with little temporal variations. Three clusters of regions with different conditions to attract business have been detected.

5.3.1 *Cluster 1: The Lagging Periphery*

The first group holds regions that exhibit weak access routes, both in terms of road networks and rail transport. Another distinctive characteristic that these regions display is their low population density, which indicates the absence of large urban centers and thus became less attractive for urban populations. In detriment of their absence of urban cores, these regions tend to lose key demographics that pursue more urbanized areas given economic and sociocultural amenities. An example is the Alentejo region where firms do not have the capacity or interest to go through with significant investments concerning R&D expenses as the entrepreneurial fabric in such regions is insufficiently dynamic and innovative.

R&D departments of large firms, on the other hand, allocate themselves in areas of higher population density, quite contrary to underdeveloped regions, whose traditional industries are stagnant or only produce intensive labor for big firms. The R&D expenses carried out by state and higher education institutions are superior to those carried out by private firms; these are, however, not significant given the absence of funding, technology, or personnel.

In terms of employment connected to R&D in the entrepreneurial sector, firms from those regions show low payment rates. Governmental sector employment related to R&D, however, shows higher payment rates. This suggests that there is either inefficiency of human resources assigned to R&D or that insufficient funds undermine more demanding research. The same occurs in higher education, where R&D-related employment is moderate. In terms of education, in its various stages, figures show that these regions maintain a low student population percentage except for the primary school sector. As a result, education tends to lag behind in these regions. The high level of children attending primary education shows that there is a shallow level of academic success, which results in students later abandoning their education. In these regions, workforce flexibility is also not very high, given the low qualifications and the lack of specialized labor skills. Indeed, the value generated by the correspondent variable seems to be a direct result of the insecurity jobs hold in these regions rather than the lack of qualifications or skills available.

Another distinctive feature displayed by these regions is the low per capita GDP, an indicator of slow development. In terms of gross added value (GAV), there is a high percentage of the primary sector, clearly revealing that these regions remain significantly behind in terms of development. Also, the GAV structure shows a proportion of the services twice the proportion in industry. This is because various regions are highly dependent on tourism, as in the case of Algarve, Madeira, and Baleares.

5.3.2 Cluster 2: The Growing Regions

Group 2 is constituted by regions that have good access routes, particularly roadways, and maintain a high population density. These regions have usually larger urban centers that attract populations from more deprived areas by offering better living conditions and more job opportunities. The R&D effort carried out in these regions is relevant, although R&D expenses practiced by the entrepreneurial sector remain significantly lower than those practiced by firms from more developed areas, those that are examined in cluster 3. It may be observed that the entrepreneurial sector from these regions is visibly becoming aware of the importance of R&D's importance. This is a sign that these regions are beginning to develop entrepreneurial firms that regard R&D as fundamental in terms of their success and competitiveness. In these regions, government and institutions of higher education replace firms that do not yet hold a robust capacity to invest in this area. R&D employment is in tune with the number of expenses carried out. In terms of education, these regions display good results concerning the number of students that attend higher education. The values for elementary and middle stages of schooling are satisfactory, without showing significant levels of failure. Workforce flexibility is average and supports the idea that the level of flexibility remains weak and needs to reach the same levels of more developed regions. In terms of per capita GDP, these regions show to be intermediary, illustrating an average level of development. Lastly, the GAV distribution shows a very weak proportion for the primary sector. A strong weight, however, for the tertiary sector (almost twice as much as the secondary sector) is patent. Additionally, evidence that GAV is nearing values from more developed regions is shown.

5.3.3 Cluster 3: The Spearheading Regions

And, finally, those regions offer good access routes, in particular railways, which exhibit less costly in the transport of goods and generate less traffic problems. As expected, France was shown to possess a higher number of motorways and shown to maintain railways as a traditional and effective form of transport and networking. However, oddly, the population density is low. Still, if we consider the fact that these regions surround Île-de-France, which bears the French capital, Paris, then this low population density results from the capital's pole of attraction. In terms of R&D expenses, these are significantly high, mainly arising from private sector. It is also due to this that most entrepreneurial companies choose the location they do, focusing on competitive product quality and innovation. There are advantages when private sector develops R&D instead of the public sector, since these are bound to be performed more efficiently and rationally in research areas of greater need. In turn, these regions are characterized according to a better qualified labor force and thus naturally bypass the need of hiring an excess of highly qualified workers as required in the previous group of regions. As such, with exception of lower secondary education, these regions present average figures for the different schooling stages and can be tied to the low failure rate in schools. These regions also seem to prefer more technical education, whereby more job specializations are available and job responsibilities delegated earlier to young people without the need to attend higher education. Supporting strong qualifications and job specialization in these regions is the high flexibility that is evident. In terms of per capita GDP, this is naturally superior, while the distribution of GAV is identical to those regions belonging to cluster 2. The small difference weighing the primary sector is related to the sparkling wines weighing heavily in the Champagne-Ardenne region.

This kind of experiment should be regularly repeated, and the determinants of attractiveness would significantly change. Noteworthy is the fact that connectivity plays a major role of attractiveness. Connectivity was identified by assuming variables related to easy transportation of people and assets. Facing a new digital area, a new study would require the introduction of networking variables such as Internet access and use of ICT by companies and small groups to better detect such groups.

5.4 Modeling Patterns of Entrepreneurial Behavior and Innovation Dynamics

As seen, space is not a static component of the development process. Its constant reshaping and its persistent change occur due to many different factors, also because of the type of relations that stakeholders generate as a result of their social and economic behaviors. Mostly, formal or not, these can be called transactions, external to the markets, internal to the hierarchies or cooperatives, and they always have an influence on the economic geography of the regions, as they are inductors of knowledge and, frequently, they also accelerate the innovative capacity of regions.

From this point of view, the introduction of time as a factor of territorial dynamics allows the transformation of simple relations between the stakeholders into lasting interactions able to create need for learning, thus learning procedures or, even, learning routines. Consequently, time can be considered a natural instrument of development. This empirical phenomenon has given origin to new very consistent theoretical approaches regarding the creation and diffusion of knowledge as a tool for the regional development process. Much important for southern Europe, this approach induces the perception of regions as historical constructions where institutions act as regulators and institutional proximity may promote development. So, there are discussions evolving from the concepts of geographic and technological proximity towards the spatial organization of agents, inducing eventual agglomeration economies. Moreover, triggered by that same interaction, very often common objectives are defined resulting in scope advantages. Hence, institutional contexts favor exchange between stakeholders.

Southern European regions have those prerequisites needed for apprenticeship from their historic-cultural past. But, just like other regions, they will also assume norms, codes of conduct, or conventions which perform functions such as research, selection, codification, transformation, control, and other procedures that, at the end, represent created or accumulated tacit or codified knowledge. From this point of view, a call for attention to the need of identifying the tools that promote knowledge such as local history, institutional governance, or stakeholders' behavioral patterns should be made.

The recognition of these arguments brings some interesting conclusions: (1) the existence of certain territories especially endowed with institutional frameworks suitable for knowledge diffusion, namely, tacit, that may explain a possible competitive advantage in their present growth path and (2) the doubt if the tacit knowledge diffusion will demand the simultaneous existence of the institutional and geographic proximity. Considering that codified knowledge is quite mobile we could ask if, when associated tacit and codified, the mobility of one does not imply the mobility of the other.

In fact, the tacit knowledge diffusion demands the need for geographic and institutional proximity associated to space, the distance being a factor that doesn't restrain the codified knowledge. Cooke et al. (2004) argue on this base the reason why only facing an institutional whole composed of scientific-technological institutions a region may be having a regional innovation system and then attain sustainable growth forms.

Furthermore, it is worth to refer that in a regional system of innovation (a territory with a cohesive entrepreneurial system particularly apt to innovate) innovating and non-innovating companies set together. The leader companies are the most responsible for the introduction of new products, differentiation, or interrelation with other firms and encouraging new technological needs through their participation in external networks with other companies (suppliers or customers) or R&D centers. Being capable of proportioning trust, they create alterations in entrepreneurial relations as well as a better tacit knowledge diffusion, which in turn eases the sources of codified knowledge transmission. However, are these alterations in the knowledge transmission processes factors sufficiently capable of promoting structural changes? In southern Europe and facing such a significant amount of European funds directed to enhance entrepreneurial activities and skills, this continues to be the most significant issue to justify the continuation of strong support system being exclusively directed to such regions (MED EU programs and other similar ones).

Eventually, an answer to the previous question may be found reflecting on the forms how cooperation takes place within the region: The theoretical contribution of Antonelli (1995) allows explaining the origin of the cooperation due to the existence of complementing among different production units, whose relations are not strictly based on competition. The functional division of the emergent economic activities and the desegregation of the productive processes encourage a major specialization degree, inducing complementariness and interdependencies.

It is the coordination that better permits the transmission of relevant information, becoming the only warranty of the productive cycle for new products. It allows maintaining a safe information flow, enabled by the ex ante cooperation, as it strengthens the concepts of learning and networking. In SMEs the learning aptitude of firms, is significantly linked to economic, social, legal, and political dimensions (Stough et al. 2007). The peripheral regions, where development is an urgent issue, deal with inflexibilities at all those levels.

Various interfaces have been created, helping firms to combine sources of technical know-how and information. Sometimes they organize institutional local networks which help creating cohesion or favorable contexts for innovation; this is certain. However, the development process of peripheral regions remains imprecise and fuzzy (Noronha Vaz and Nijkamp 2008), probably because a major driver has escaped for a long time to the several theoretical approaches. It seems inevitable to include how the learning process at local level takes place even if also in this case literature remains ambiguous due to the complexity of learning. The possibility of knowledge bases enlargement is illustrated in Fig. 5.1.

The model proposed below highlights the role of firm in the regional development process of locals and proposes interactions between firms' capacity to innovate and regional development levels. The interactive nature of the model allows accepting the complex nature of the innovation process and retaining a certain tangible feature.

When identifying, classifying, quantifying, and modelling the determinants of innovation, the local capacity of innovation may become more clear as generated at the level of a simple activity branch that may tend to more complex technological regimes of cooperation or as simpler forms of regional growth in which companies could choose to coordinate into more complex inter-territorial systems of institutional cooperation, defined along the horizontal and vertical axles of the model represented in Fig. 5.2.

Each innovative process pulls the firm to a positioning that moves continuously to the top and the right side of figure, changing accordingly the combination of determinants for future innovative choices. In the model, four major sets of innovation determinants are represented: local development conditions, technological learning, institutional proximity, and firm strategy.

To illustrate the defined model, and to the study cases presented in this chapter, a new analysis was carried out for the 11 regions across Europe listed in Table 5.1, knowing that they all represent predominantly rural, peripheral regions. While lagging, these regions show some contrast in performance in terms of economic growth, as earlier observed. Each region is relatively homogeneous internally in terms of its econonomic activities and social relations. Their discrete administrative

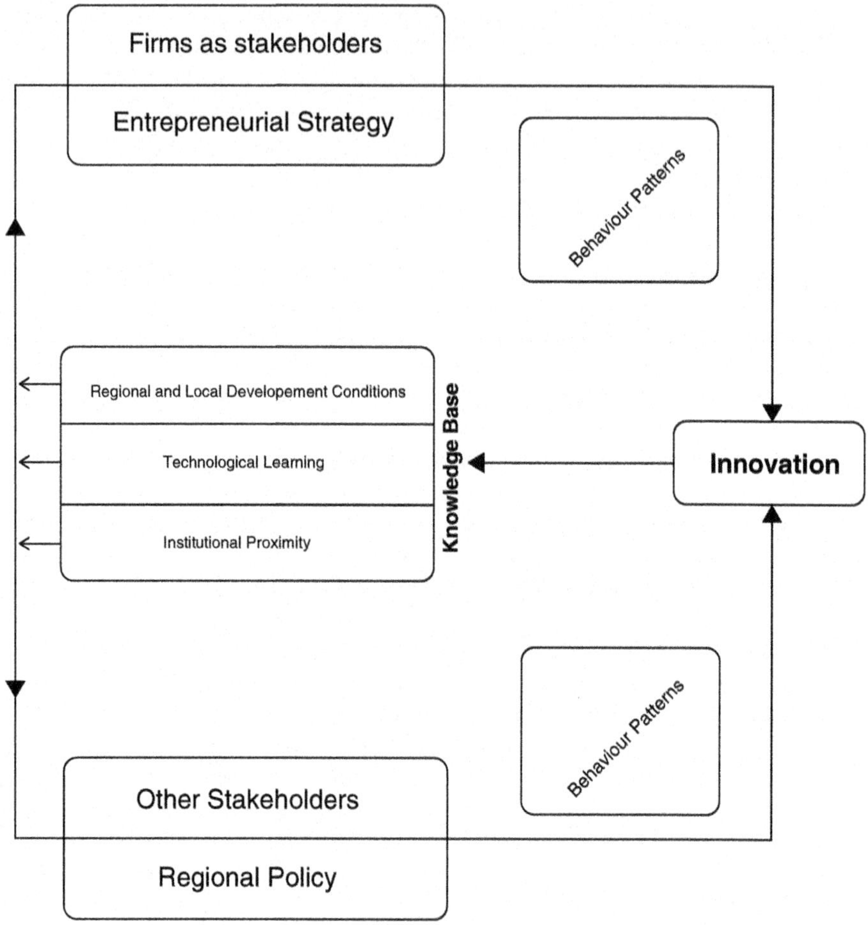

Fig. 5.1 Firms' innovative behavior and knowledge base enlargement
Source: Noronha Vaz and Cesário (2008)

borders (approximating NUTS II or III regions) facilitated the collection of secondary data. After the presented territorial systems, a larger set of statistical information became available from secondary sources allowing subsequent quantitative analysis such as statistics on the demographic structure, as well as the data on the economic and social performance. Assembled data for 17 indicators reflected aspects of the local capacity for growth, as well as the financial support structure available to promote regional development based on the capacity of local firms to innovate.

A detailed questionnaire with mainly closed questions provided materials for analyses of the following aspects, as in Fig. 5.2: the overview of the firms; the characteristics of the top manager; the history and the profile of the firm; the quality of manpower and training; the products and processes and respective changes; the

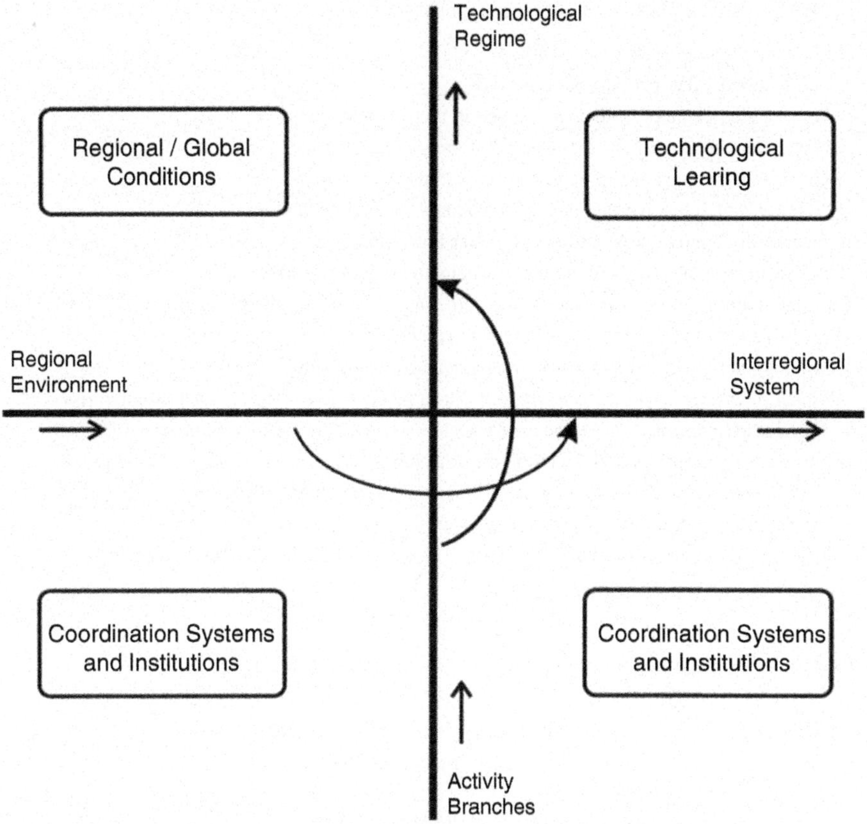

Fig. 5.2 Firms' innovative behavior and knowledge base enlargement
Source: Noronha Vaz and Cesário (2008)

inter-company relationships; and the relationships with support bodies and other aspects of the regional environment.

A full-scale survey involving a sample of 323 firms and different variables related in the previous figure (namely, variables of entrepreneurial strategy, variables of coordination systems and institutions, variables of technological learning and regional conditions) served to detect innovative behaviors across regions. See Table 5.1.

For multiple innovators, see Table 5.2; the study detects that two variables of local and historical nature are significantly associated with product innovation: regional household expenditures on drink and the percentage of industrial employment in the total regional employment. Contrarily to what is the common assumption, the inverse correlated variable to product innovation in these regions is the duration of the top manager in the firm. The complexity of multiple innovators seems not to be conforming with long-time managers, probably because those are the ones that have been less exposed to external learning forms. However, in the case of

Table 5.1 Variables indicative of regional global conditions

I Internal regional product per capita
G Gross added value per person employed
G Gross added value per person employed in the primary sector
Gross added value per person employed in the secondary sector
G Gross added value per person employed in the tertiary sector
P Proportion of active population with secondary level qualifications
P Proportion of active population with higher level qualifications
Investments in transports and communications in relation to surface area
E Employment in the primary sector as a percentage of total employment
E Employment in the secondary sector as a percentage of total employment
E Employment in the tertiary sector as a percentage of total employment
Number of enterprises in the primary sector as a percentage of total number of enterprises
N Number of enterprises in the secondary sectors as a percentage of total number of firms
N Number of enterprises in the tertiary sector as percentage of total number of enterprises
E Expenditure on food as a proportion of household expenditure
E Expenditure on drink as a proportion of household expenditure
E Expenditure on meals outside the home as a proportion of household expenditure

Source: Noronha Vaz and Cesário (2008)

Table 5.2 The determinants of innovation for multiple innovators

Dependent variable	Exogenous variables by group of determinants	St. coeff	t	Sig.
Product innovation	Years the top manager enters this enterprise initially	−0.258	−3.655	0.000
	Household expenses in drinks/household expenses total	0.268	3.799	0.000
	% employment in industry	0.167	2.437	0.016
Organizational innovation	IRP per capita	−0.302	−4.549	0.000
	% of turnover spend on R and D	0.224	3.392	0.001
	External factors region – similar enterprise	−0.332	−4.691	0.000
	External factors – overall	0.445	5.868	0.000
	Years the top manager enters this enterprise initially	−0.133	−1.998	0.047
	External factors – customers	−0.195	−2.694	0.008
	Household expenses in catering/household expenses total	−0.165	−2.347	0.020
	Governmental assistance: national	0.132	1.988	0.048
Marketing innovation	Household expenses in drinks/household expenses total	0.347	5.617	0.000
	New services/total services	0.205	3.343	0.001
	Intellectual property	0.210	3.346	0.001
	External factor region – research institution	−0.193	−3.098	0.002
	External factors – overall	0.490	5.355	0.000
	External factors – research institution	−0.343	−3.739	0.000
	External factor region – equipment suppliers	−0.156	−2.484	0.014

Source: Noronha Vaz and Cesário (2008)

non-innovators, product innovation is linked to other different factors such as the variables of local and historical nature which are negatively correlated to industrial employment and productivity in services. The learning factor influences positively the firms as a qualification of the top manager in business and economics or as a turnover spent by the firm in R&D.

Further comparison of results shows that the determinants of innovation, in both multiple innovators and non-innovators, for organizational innovation to take place demand from non-innovators efforts that are mainly related with external conditions – either related to the level of local growth (internal regional product or percentage of the employment in services) or related to institutional proximity and learning (contact with research institutions or training). In the case of multiple innovators, the set of determinant factors is much larger.

Again, a long-staying manager does not contribute positively to innovativeness, and the local and historical conditions that may further restrain it are related with the existence of similar firms in the region or increases in the regional product, accentuated by increases in the expenses in catering. All the external relations excepting those with customers are positive determinants of organizational innovation indicating the importance of technical proximity for this group. Learning and investing are the other components to be considered in the percentage of the turnover spent in R&D as well as support from governmental assistance.

Marketing innovation benefits mainly from general local development conditions in the case of multiple innovators, while non-innovators need to include in their efforts close contact to research institutions in presence of significant labor productivity in services to have it. In the first case, equipment suppliers pull marketing innovation; in the second one, the costumers promote it. Both patents and technical qualifications are determinants that must be considered in the case of non-innovators (Table 5.3).

Those firms that carried out mainly one kind of innovation have been classified as focused innovators. Despite the fact that these firms orient innovation mostly to marketing and organizational forms, we have detected two very different performance levels. Those firms with low performance that still produce new products seem to find incentives in regional highly qualified labor force, in intellectual property, and in a better qualification of the general manager, representing a clear focus on parameters related to learning within the region, as shown in Table 5.4. The same dependency from learning, necessarily internal to the region, is confirmed when regarding organizational innovation. It results from the percentage of firm's turnover spent on R&D and from the interactions with external factors in general (belonging to groups of technological interest outside of the region).

The high-performance focused innovators do not integrate, in general, product innovation. As demonstrated in Table 5.5, they concentrate their innovativeness in the marketing or organizational processes. In these cases, firms' dimension matters

Table 5.3 The determinants of innovation for non-innovators

Dependent variable	Exogenous variables by group of determinants	St. coeff	T	Sig.
Product innovation	GAV services/employment in services	0.979	9.732	0.000
	% employment	−0.321	−3.732	0.000
	% of turnover spend on R&D	0.193	2.583	0.012
	GAV services/employment in services	−0.304	−3.595	0.001
	Educational qualification of the TM – college certificate	0.253	3.242	0.002
	External factor region – equipment suppliers	−0.184	−2.236	0.028
Process innovation	External factors – equipment suppliers	0.282	2.502	0.014
	GAV services/employment in services	0.316	3.920	0.000
	External factors – overall	0.290	2.540	0.013
	Years did the top manager enter this enterprise initially	0.200	2.481	0.015
	LFQ secondary/active population	−0.183	−2.299	0.024
Organizational innovation	External factors – research institution	0.393	4.257	0.000
	Training carry out – types	0.301	3.194	0.002
	IRP per capita	−0.351	−3.207	0.002
	% employment in services	0.234	2.120	0.037
Marketing innovation	GAV services/employment in services	0.840	7.316	0.000
	External factors – customers	0.230	2.782	0.007
	% employment in industry	−0.267	−2.954	0.004
	% of turnover spend on R&D	0.205	2.702	0.009
	GAV industry/employment in industry	−0.266	−2.775	0.007
	Intellectual property	0.185	2.445	0.017
	% qualified technical	−0.169	−2.147	0.035

Source: Noronha Vaz and Cesário (2008)

Table 5.4 The determinants of innovation for focused innovators with low performance

Dependent variable	Exogenous variables	St. coeff	T	Sig.
Product innovation	LFQ high/active population	0.737	6.704	0.000
	Intellectual property	0.379	3.410	0.002
	Area of qualification of the TM: business/economics	0.337	3.085	0.005
Organizational innovation	% of turnover spend on R&D	0.543	4.887	0.000
	External factors – overall	0.443	3.981	0.000
Marketing innovation	GAV/total employment	−0.414	−2.449	0.021

Source: Noronha Vaz and Cesário (2008)

in order to allow the introduction of new tasks and their reordering or specialization. This may be the reason why we also have detected that information and technology specialists constitute another source for innovative initiatives in this group as well as

Table 5.5 The determinants of innovation for focused innovators with high performance

Dependent variable	Exogenous variables	St. coeff	t	Sig.
Organizational innovation	People normally work in the business	0.608	4.985	0.000
	New services/total services	−0.544	−4.165	0.000
	New industry/total industry	−0.429	−3.402	0.002
	External factors region – IT specialists	0.322	2.728	0.011
	Training carry out	0.229	2.125	0.043
Marketing innovation	People normally work in the business in 2000	0.478	3.406	0.002
	LFQ sec/active pop 1997	0.454	3.234	0.003

Source: Noronha Vaz and Cesário (2008)

the training carried out of the firm. Such competences are acquired externally to the region. It was possible to have a reasonable understanding from the mechanism behind marketing innovation, particularly in what concerns focused innovators. The number of workers and their qualification played an important role to justify it.

Comparing the factors for innovation between multiple innovators and focused innovators, the complexity of the model for the first group confirms. Product innovation that can only be explained by the group that has lower performance levels results mainly from two factors directly related to the firms and to the qualifications of the labor force, the latter related to local development variables. The same can be observed for organizational innovation or marketing innovation, for which a simple combination of factors is linked directly or indirectly to firms' strategic choices.

5.5 Useful Conclusions for Southern Europe

All together the applied methodology demonstrated the irregularity of cause-effect conditions to generate innovation. The fact that in a relatively small sample of European firms, it was possible to detect, at least, four different behavioral patterns towards innovation explained by inconstant factors, mostly of microeconomic nature, is important as well. Surprisingly, the number of years that the general manager has been in the firm and firms' size (in a few cases) were the only detectable variables related to firm strategy (determinants exclusively internal to the firms) with clear impact over innovation. All the other significant variables are closely related with local development conditions, technological learning, and institutional proximity, confirming the importance of our model.

Our conclusion follows: Although firm strategic choices determine innovation, the microeconomic base is clearly insufficient to explain the aptitude of the firm to produce new products and processes. Variables related to technological learning appear to play a direct role in the increase of new innovative forms in the small firms. Percentage of the turnover expended in the R&D activities, training carried out in the firms, technical qualification of the top managers, and intellectual property are to be considered as the most important factors for all the different behavioral patterns in peripheries. Institutional proximity is associated with technological learning. Even so, we have detected some specific inputs related with requirements that both suppliers and clients impose on small firms that led them to innovate, particularly when we deal with firms that are characterized by difficulties in processing innovativeness as it is the case of non-innovators. In general, all the groups profit from generalized external exchanges in most of the different innovative forms. In southern Europe, one of the regions with such characteristics, innovation needs additionally to microeconomic conditions a favorable meso-economic context.

Macroeconomic conditions also promote firms' innovativeness aptitude: Factors related to regional growth like gross added values created by services, household expenses, or labor force qualification associate with all forms of innovation, independently of the behavioral pattern followed by the firms. Still, these variables have puzzling effects. In many cases, direct effects do not occur as expected showing inconsistent relations between regional growth, consumption or market competition, and innovation. Another important conclusion is the idea of direct financing to firms or exclusive support of information technologies. The research indicates that the only impact of such measures occurs on marketing innovation and this is only in the case of good performers both multiple and focused innovators.

One of the most important conclusions, however, that needs to be here emphasized is a similar kind of very restrictive positive impact that arises when considering any kind of governmental assistance: A reduced positive impact, related to marketing innovation, was observed just in the case of multiple innovators. Also, a major conclusion for regions is the fact that for the most dynamic group of companies, internal consumption is a determinant factor to generate product innovation, while institutional proximity and networking are driving forces for organizational innovation. This last conclusion has a major impact of how business chooses regions and how they tend to search for the best place to locate when looking for partnerships and the best ecosystems. We should note that when considering the specificities of small businesses, the above-described arguments gain more relevance. Unlike big firms, the small business interacts intensely with the territory in which it locates, as a sign of embeddedness. The particularly tight links developed with the external environment also reduce uncertainty risks. In general, they do not only locate near the residence of their owners but also have geographical and sociological proximities as their main sources of assets and information.

In terms of policy suggestions, authors have been interested in knowing if the recent regional policies based upon innovation strategies for smart regional specialization will bring progress and sustainability to southern Europe and several studies point out that results are very variable from effective to indifferent, depending on the region and period studied. Some authors have defended that tailor-made policies would be more consequent for such regions; others assume that the existing funding brought significant improvements with scientific outcomes but with an end effect of very sporadic impacts on growth and development. And regarding the imposition for a strategic plan prior to RIS3 support may very well be an active leverage to help behavioral changes in the future, although its acceptance related to the regional forces for and against change will ultimately decide for regional success or regional failure.

References

Antonelli C (1995) Économie des Réseaux: variété et compléntarité. In: Ralllet A, Torre A (eds) Économie Industrielle et Économie Spatiale. Economica, Paris, pp 253–272

Becattini G, Rullani E (1995) Système local et marché global: le district industrial. In: Ralllet A, Torre A (eds) Économie Industrielle et Économie Spatiale. Economica, Paris, pp 173–192

Cesário M, Vaz MTN (2011) The effects of globalisation on the performance of labour-intensive industries from southern Europe: the role of localised capabilities. Int J Econ Manag Sci 1(4): 53–63, ISSN 2162-6359

Cooke P, Braczyk HI, Heidenreich M (eds) (2004) Regional innovation systems: the role of governances in a globalized world, 2nd edn. Routledge, London

Eliasson G et al (1990) The knowledge based information economy. Industrial Institute for Economic and Social Research, Stockholm

Maskell P, Malmberg A (2006) Localised learning Revisited. Camb J Econ 37(1):1–18

Noronha Vaz T, Cesário M (2008) Driving forces for innovation: are they measurable? Int J Foresight Innov Policy 4(1–2):30–50

Noronha Vaz T, Nijkamp P (2008) Large-scale production and market segmentation: an uneasy relationship. In: Vaz MTN, Nijkamp P, Rastoin JL (eds) Traditional food production facing sustainability: a European challenge. Routledge, London

Noronha Vaz T, Barbosa A, Cesário M, Guerreiro A (2003) Regional attractability to business. An empirical application to southern European regions. New MEDIT J (3/2003):52–57 University of Bologna, Itália

Scott AJ (2006) Creative cities: conceptual issues and policy questions. J Urban Aff 28(1):1–17

Storper M, Scott AJ (2009) Rethinking human capital. Creat Urban Growth J Econ Geogr 9(2):147–167

Stough R, Nijkamp P, Noronha Vaz T (2007) Local knowledge and innovation policy. Environ Plann C 25(5):633–637

Tsipouri LJ (2017) Innovation policy in Southern Europe: smart specialization versus path dependence in advances in the theory and practice of smart specialization. In: Radosevic S, Curaj A, Gheorghiu R, Andreescu L, Wade I (eds) . Academic Press, London, pp 125–155

Vaz E, de Noronha T, Pinto H (2018) Conclusion: resilience—what's next? In: Pinto H, Noronha T, Vaz E (eds) Resilience and regional dynamics. Advances in spatial science, The regional science series. Springer, Cham

Chapter 6
Modelling Regional Innovation Patterns: The Case Study of Portugal

Abstract Evolving from the discussions of the previous chapters herewith and still taken by the understanding of what innovation is about, new methods are presented that allow to emphasize regional innovation structures and depict the entrepreneurial micro-economic view of the firm at the local level. These proposed methods are of interest to regions in need of a long-lasting accompaniment of public policy actions. By relating geographic proximity and clustering, the best factors for regional growth are detected. The methods described in the present chapter are mostly adequate for lagging areas where investments and incentives are planned to occur, either utilizing local development targets or business promotion, such as southern Europe. We try to emphasize how modeling the specific regional innovation patterns at the local or regional level can be a significant aim to help make more fruitful investment decisions.

Keywords Regional innovation · Innovation modelling · Entrepreneurial behaviors · Lagging areas · Southern Europe · Regional sustainability

6.1 Regional Economic Dynamics

The previously used methodologies brought some light into the processes of measuring and modelling the drivers and impacts of innovation. Still, new methods have been developed that are able to emphasize the regional aspect of innovation not restricting thereby the entrepreneurial microeconomic view of the firm. Some of these methods have been developed on a basis of concepts related to geographical proximity and clustering such as knowledge as key factor of growth, knowledge spillovers as limits for location, complementarities as advantages of knowledge diffusion, knowledge filter as key advance in the theory of entrepreneurship, and institutional structures of innovation and innovation pathways as recent contributions to innovation modelling (Galindo et al. 2011; Vaz et al. 2012a, b).

In a case study that observes three Portuguese regions (one central and two peripheral ones) and their institutions, included in a web-based inventory of six hundred Portuguese innovation institutions, a dataset was analyzed by means of

E. Vaz, T. de Noronha, *Sustainable Development in Southern Europe*,
https://doi.org/10.1007/978-3-662-62177-6_6

principal coordinates analysis followed by a logistic biplot approach to design a systemic typology of innovation structures where each institution is individually represented. In this book, the details of the applied methodology will be set aside. However, it should be mentioned that the institutions were randomly selected from Internet sites using Portuguese keywords directly related to innovation; all the web-published explicit descriptions concerning innovation were investigated and content analyses applied and codified into empirical attributes such as knowledge promotion, strategic management, R&D promotion, knowledge transfer, partnership and cooperation support and governmental orientation, and skills development. Principal coordinates analysis and a logistic biplot application allowed an exact classification of innovation profiles submitted to Voronoi diagrams. Aiming to investigate the contribution of knowledge to regional prosperity, institutional innovation profiles and regional innovation patterns have been traced in that study. The goal was to be able to address the following research question: "Due to different absorption capacities or different knowledge filters institutions have innovation individual profiles. Is it possible to observe each one of these profiles, map them and report them to a set of profiles of other nearby located companies (in the same country, region or cluster)? What can such a static comparative analysis tell us?" (Fisher 2006).

Further contribution to the modelling perspective is considerably enriched by a new approach of models of urban growth, adding up the dynamics of regional change. The fragmentation of landscape and irreversible consequences on land use and ecosystems has become an increasing concern worldwide (Nagendra et al. 2004). Brought by economic growth, the creation of new infrastructures to support demand has led to profound structural changes in the traditional concept of city (Davidson 1998), changing the local perception of city to a larger urban agglomerate, understood as the urban region. While cities represent traditionally a positive source of economic prosperity, the agglomerating city (Rosenthal and Strange 2001) sets out a new role on the importance of urban areas and highlights the concern if the options of sustainable urban growth are the best we can do for future generations and faces new challenges such as promoting a low-carbon society while promoting the vitality of the urban-rural fringe (Han et al. 2012). Pollution, for instance, brings an adverse impact to development, and most of the urban regions today show an augmenting number of mortality brought by pollution as well as territorial occupation.

Cities have also brought an increasing criminality rate, resulting from an excessive population concentration and inability to cope with excessive concentration of people in urban areas (Cusimano et al. 2010). Nevertheless, cities still are a source of hope for better lives, and with growing enhancements at a technological level, the existence of areas is hosting the possibilities of creating better transportation (Litman and Burwell 2006), health, and education systems, as well as fostering social movements (Nicholls 2008) leading to a better society through the existence of urban concentrations as long as well managed and driving with sustainable solutions. The eco-friendly city has brought greener cities to a research agenda of environmental science and offered as a consequence a holistic vision of urban processes

(Roseland 1997), where the interactions of humans with ecosystems can promote a better social conscience, but also lead to a clearer understanding of how urban regions must interact with nature. One of the most important interactions is linked to the carrying capacity of land, in which a function analysis, function valuation, and conflict analysis should be present to understand the landscape dynamics and foster sustainable development (de Groot 2006). This is strongly linked to a better understanding of the spatial dimension of land use change and land use transitions, fostered by the integration of geographic information systems as tools to assess and manage present and future change (Vaz et al. 2012a, b). Human being, as an economic being, is reshaping its environment neglecting its role as a keystone species (O'Neill and Kahn 2000) however the impacts of unforeseen, and often unpredictable, consequences of its interaction with the world. In this sense, land use plays a vital and fundamental role in maintaining a balance on the environment in general, and one of the culprits of urban growth has become the loss of certain types of land that have a unique function for the complex interactions of the environment (Xiao et al. 2006).

An entire set of literature has been built on this premise concerning impacts of land use and benefited greatly a more accurate and better planning for sustainable development at a spatial level (Batty 2005). These models are designated by urban growth models and draw largely from an integration of geography, economics, mathematics, and social sciences. The possibility to manipulate spatial data and visualize the cities of tomorrow at spatial level has advances with the computational capabilities through the field of geographic information systems that, much more than a tool to edit and query certain features in space, has proven to be a tool to understand underlying patterns of spatial phenomena, offering through spatial analysis a quantifiable and assertive technique to propose a better understanding of the spatial structure. The scattered nature and the fragmented characteristics of urban growth are however hard to fully grasp. While land use change models and urban growth models foment stochastic understanding of possible outcomes of urbanization processes, no consensus exists on which variables play a distinctive role on urban change. This is largely a result of the different types of urban regions and the diverse policy agenda that countries may have. Coping with urban change, however, is a common prerequisite to maintain our quality of life at present. While in the short term local sustainability plays a major relevance, in the long term common structural changes must also be considered.

6.2 Spatial Aggregation and Geometrical Clusters

The debate on the spatial clustering phenomena requires the concept of regional innovation systems (RIS) – defined as a network of organizations, institutions, and individuals, within which the creation, dissemination, and exploitation of new knowledge and innovation occur (Cooke et al. 2004). The link between "clusters" and RIS may be explained by the fact that within some spatial systems, groups of similar and related companies (e.g., large and small companies, suppliers, service providers, customers, rivals, etc.) comprise the core of the cluster, while academic and research organizations, policy institutions, government authorities, financial actors, and various institutions for collaboration and networks make up the innovation system of which the cluster is a part (Teigland and Schenkel 2006). It has been shown (Arthurs et al. 2009) that the patterns of close and remote relationships (including those taking place within a cluster) vary, at least, by industry, ownership status, and market orientation, as well as in conformity with the growth phase and size of the cluster. This kind of system occurs, independently of form of relationships or cluster dimensions (Davis 2008); also in peripheries different structures of interaction and different innovation pathways can be detected (Monteiro et al. 2013).

The graphical representation of the regional determinants of innovation in Portugal shows three very contrasting results for each of the different considered regions. When considering the relation of the variables/attributes to the innovation gradient, we can conclude that, for Portugal, in general, the attributes "promoting knowledge," "managing," "promoting R&D," "knowledge transfer," "promoting partnership and cooperation," and "orientation" are the most influential ones. Per region, we can evaluate the importance of each attribute for the set of institutions, thus supplying material for regional development policy considerations. The application of the biplot to the Portuguese regional scene also confirmed that in those cases of higher institutional innovation, a greater variety of attributes was being used. Not all the attributes are used with the same intensity, or they are not easily available. For some reasons, institutions are not able to absorb them, or there is a different elasticity for each attribute – this topic is asking for further investigation.

By detecting the types of structures underlying the institutions in Portugal, many advantages and fragilities may be identified and clearly interpreted, both from a micro- and a macroeconomic view. For Portuguese policy makers, some lessons can be derived, such as total geographical asymmetric use of attributes by institutions and massive concentrations of the most innovative performances in the Lisbon and Porto areas. The reasons to justify such contrasts may be identified at cluster level, by region, and the solutions should arrive after detailed individual institutional profile analyses and application of specific actions – so much for the country case (Figs. 6.1, 6.2 and 6.3).

Also, managers and other executives in companies or other institutions can compare their individual profiles, reproduced in a geometrical location, with that of the system average using a useful tool to reinforce specific measures and improve the relative positioning – this may be done by seeking a more intensive use of the missing attributes. Finally, this method provides a systematic empirical basis for a solid and informed discussion on regional cluster architecture to help focus policies for regional development.

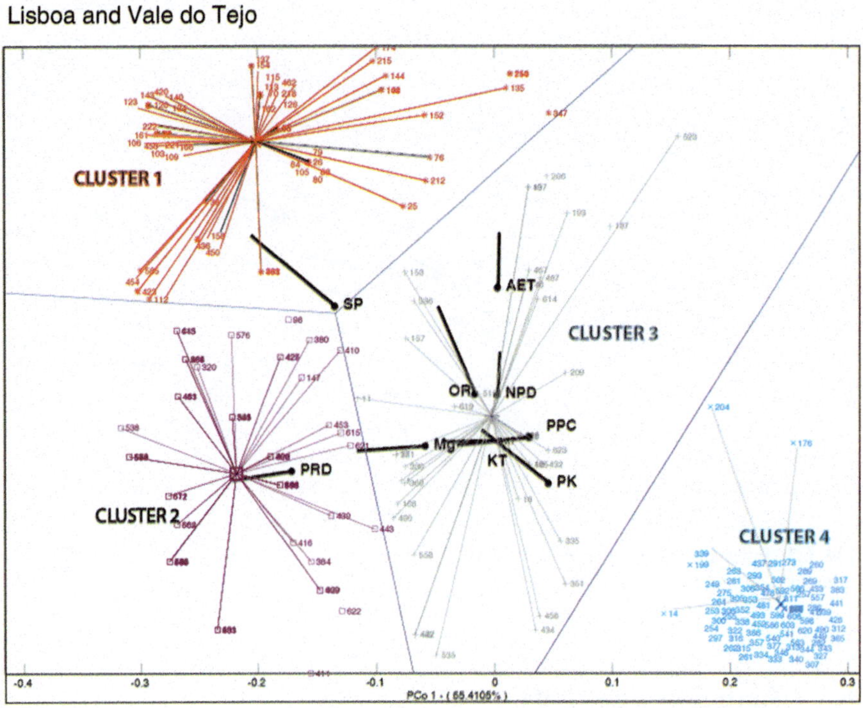

Fig. 6.1 Structures of innovation in Lisboa and Vale do Tejo and Norte
Source: Noronha Vaz et al. (2013)

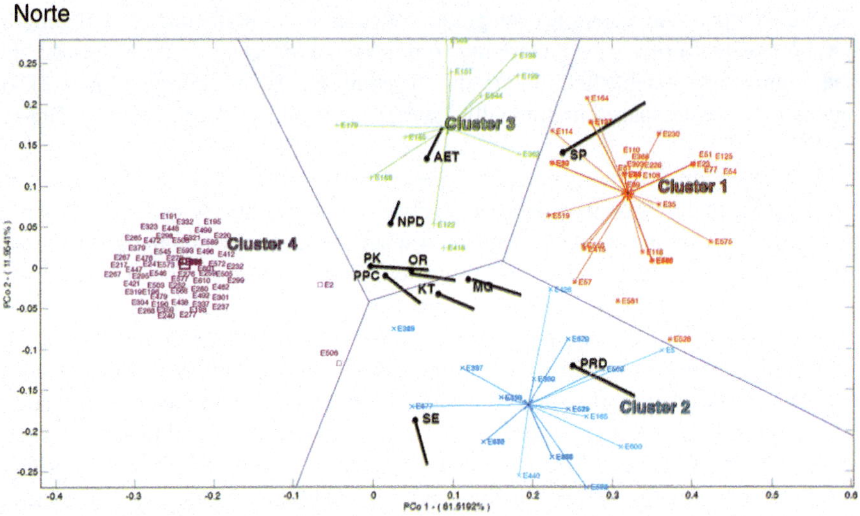

Fig. 6.2 Structures of innovation in Centro
Source: Noronha Vaz et al. (2013)

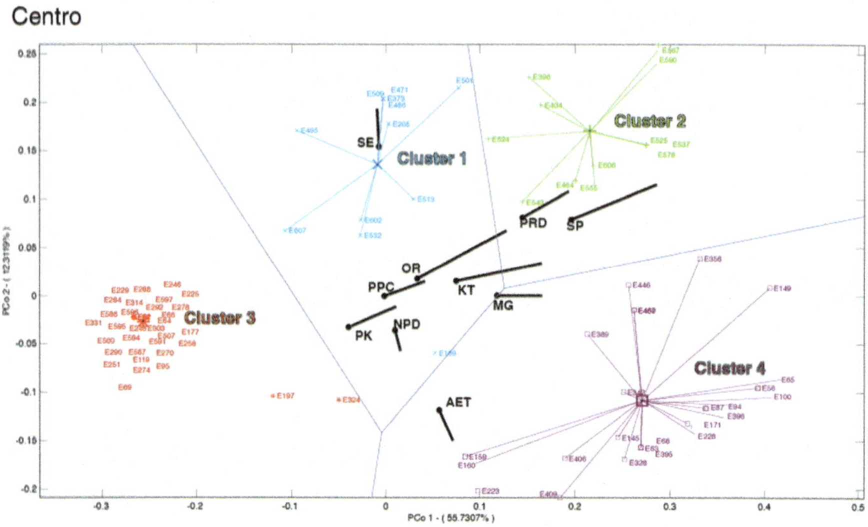

Fig. 6.3 International Innovation networks
Source: Noronha Vaz et al. (2013)

6.3 Spatial Connectivity

Spatial information has enabled the possibility to understand the relations over space of different types of features (Jankowski 1995). The spatial properties of location of activities and respective impacts are still far from being completely understood and have developed into a complex integration of economics, mathematics, and geography. A reason for this is the underlying complexity of the spatial patterns formed and the connectivity established among the different agents in a complex network of interactions over space, traditionally studies in ecology (Moilanen and Hanski 2002).

The possibility to merge the configuration of features with networks may be assessed elegantly through generating a network which connects spatial information of features. The connectivity of features in space allows understanding and fostering the dynamics of collaborations of innovation from a spatial perspective. This was achieved by converting the provided street addresses of the businesses into a point vector in space. The address is categorized into its locational determinants entailing its street number, street name, and postal code. This was then added into ArcGIS 10.1 where the process of spatial connectivity – correspondent to the transformation of the address into a point – was carried out. The geocoded addresses were than exported into Google Earth, to match the consistency of the location through attribute properties of the surrounding area, as well as confirmation of metadata related to the geocoded feature.

In our precise case, all the institutions belonging to cluster 1, assumed to be the most innovative one, were investigated, and the respective links reported till the fifth connection – considered at any geographical level (local, national, or international). Because several institutions had no reported links, the sample that was used for the mapping was reduced to 37 institutions in a total of 65 point features. The point features allowed to create a total of fifteen groups. These groups of points were then connected by relevance of indicated partners, allowing establishing a spatial understanding of small networks with spatial connectivity. The points, thus, were then converted into line segments and projected accordingly on the map.

Figures 6.4, 6.5 and 6.6 defines the connections found at different scales, global, national, and local, and report to the 50 most innovative institutions in Portugal, all included in cluster 1 and considered to be the most innovative in the country. Only a few relations are pointed out to exist between the spatial component of countries and business innovators. In fact, most of the relations even at national level are formed only above the Tejo valley, being Lisbon and Porto the main hubs for partnerships.

By detecting the types of patterns of structures of innovation in Portugal, many advantages and fragilities may be identified and clearly interpreted from a meso-economic perspective:

1. Spatial connectivity delivers a combined method able to evaluate the kind of connections underlining the innovation taking place at a certain region or country.
2. In our particular case – the application for Portugal – we can confirm the asymmetric flow distribution resulting from the connections from the most innovative institutions, which have based their innovation above all on the

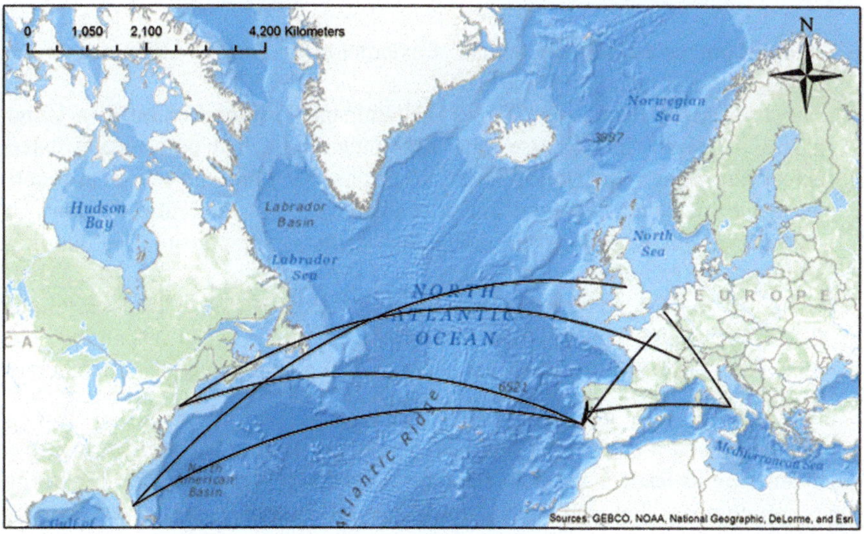

Fig. 6.4 Flow design for international connections

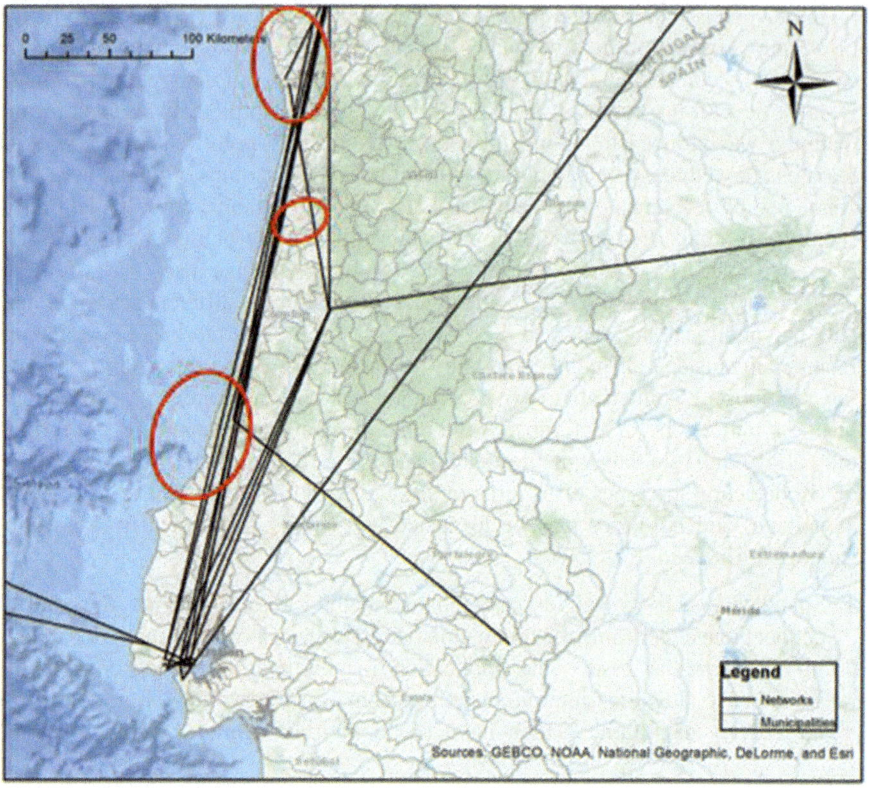

Fig. 6.5 Flow design for internal connections in Portugal

study of processes (SP), on the use of external technologies (AET), and on new product development (NPD).

3. The asymmetric distribution shows predominant flows concentrated in Lisbon area and Oporto (in this case much less intensively) that occasionally extend across Europe or to the United States. When observing the connections at country level, we may find two hubs and a small focal point in Centro Region. The method permits to pick up the individual institution responsible for this flow, searching for its innovative prospects.
4. Contrarily to what was expected, not many connections start in the same point in the Lisbon region. This indicates that different institutions can sustain their own innovation paths in a structure that although still not very complex or elaborated defines interconnections at an elaborated level.

The addressed model offers multiple advantages to access the performance of companies by its leaders and policy makers. Leaders of companies or other institutions can compare their individual profiles, reproduced in a geometrical location, with that of the system average by using a simple tool, concluding whether or not they should reinforce specific measures to improve their relative positioning – this

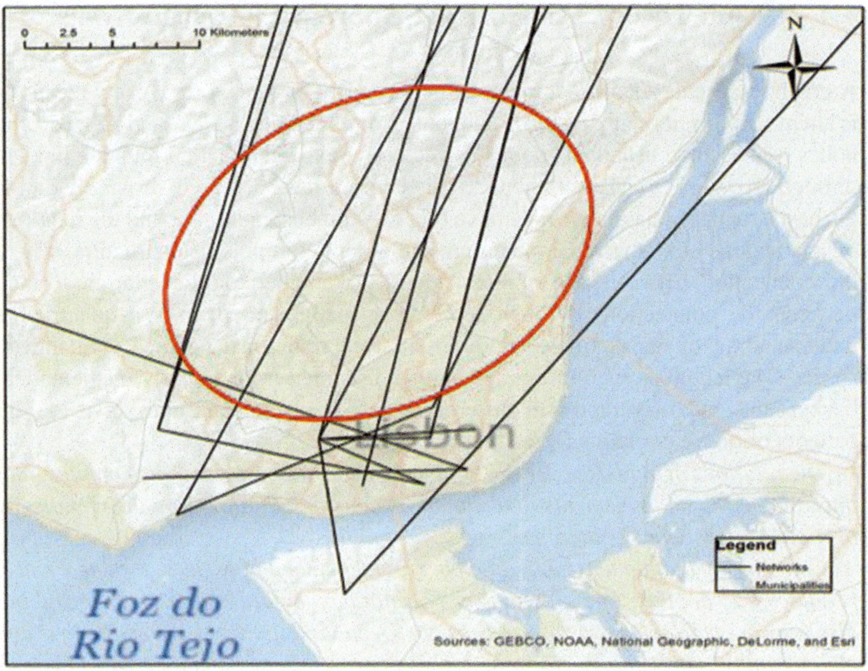

Fig. 6.6 Details of connections in Lisbon area

may be done by looking for a more rigorous use of the missing attributes, for example. Also, for policy makers and planners, this model could become a powerful tool. As pointed out, this study confirms the need to implement tailor-made policies to endorse innovation at regional level. Such is only possible when identifying the specific choice of attributes used by the set of companies and other institutions. The pattern they define to innovate may suggest those specific measures required to act directly on each described attribute contributing to a new concept of intervention – the regional cluster architecture, to help focus policies for regional development.

The examination of flow designs recommends that the emergence of innovation is also a result of the flow intensity which submits the innovation processes as a spatial determinant. Therefore, major general policies to promote it will not be able to be entirely efficient if flow design is not considered. Resulting paths should be able to create some sort of path dependency; in this case, the efficiency of promoting policies in such environments should tend to increase. The contrary is to expect when no flow design emerges in the regions.

6.4 Graph Theory: An Abstract Approach to Urban Areas

A graph represents an abstract mathematical concept of a network. By any means, it is always a portrait of any types of relations. Combined by "nodes" and "edges," the nodes (represented usually as points or vectors) are the end that permits the linkage through an "edge" to another "node." This concept has been used very early and is the best structural relation we may have of a two-dimensional space and the relations of interactions over space. Graphs represent a defined topological structure, where the connection between one of the "nodes" to another, or the relation formed between the connection, is much more than spatial and permits a quantifiable understanding of connections, far different from traditional geometric Euclidean spaces. The length as such of a graph can be discarded from a purely mathematical background and may represent different types of phenomena and serve to find patterns of discrete relations over space.

The relations of land use over the total landscape are however quite complex, and thus, one of the main elements that should be considered in framing the complexity of spatial interactions is land use.

*Considering that land use and land cover, independent of its area, format, or size, is adjacent to another type of land use. This not only makes land use connected but also harbors characteristics of similarity between land use types, given the first law of geography, where things closer to each other are inherently more related (*Tobler 1970*). This concept of neighboring land and spatial proximity is translated by a graph of a finite number of connections, varying on the size of the spatial area, type of land, and geomorphological characteristics of the terrain and the landscape through adjacency* (Fig. 6.2).

Noting that each land use may be represented geometrically as a polygon, of which a centroid can be derived:

$$C = \frac{x_1 + x_2 + \ldots + x_3}{k} \tag{6.1}$$

This centroid may thus represent the central point of equidistance within a given land use type. The geometric point that this centroid represents is characterized by centroids derived from the neighboring land use classes; thus the cumulative land use classes could be represented as $\varepsilon = c_1 + c_2 + \ldots + c_3$ *equally in* $\mathbb{R}^n$*, defined as the centroid geometrical space. This space* ε *will now be considered* $\varepsilon = G$*, where G is a pair* $G = (V, E)$ *in which V is a finite set called the vectors of G and E is a subset of V designated as edges. In the geometrical space of centroids* ε*, it is thus considered that E can be defined as long as* $V = c_n$*. I have thus transformed a geometrical space into a representation of a simple graph. Within this representation, it can be stated that the spatial relations found between the geometric spaces of land use may be represented as a simple graph, where interactions of land use can* *be processed by their adjacency to different land use types. These relations are explored visually by the definition that a graph may be force directed based on the number of connections existing within different land use types within the geographic area. This pertains to the different connections allowed between land use for the adjacency of existing land use types in a city or urban regions. The position of nodes and the location of the abstract interpretation of land use per node is of great importance for the geovisual analysis. This allows to visually enhance the size of the edges as to allow as few crossing edges among the different nodes. In the case of land use, this is of great importance as this allows understanding the relations of different land uses among the area. Thus, the ideal distance becomes proportional to the length of the shortest path between them* (Fig. 6.7).

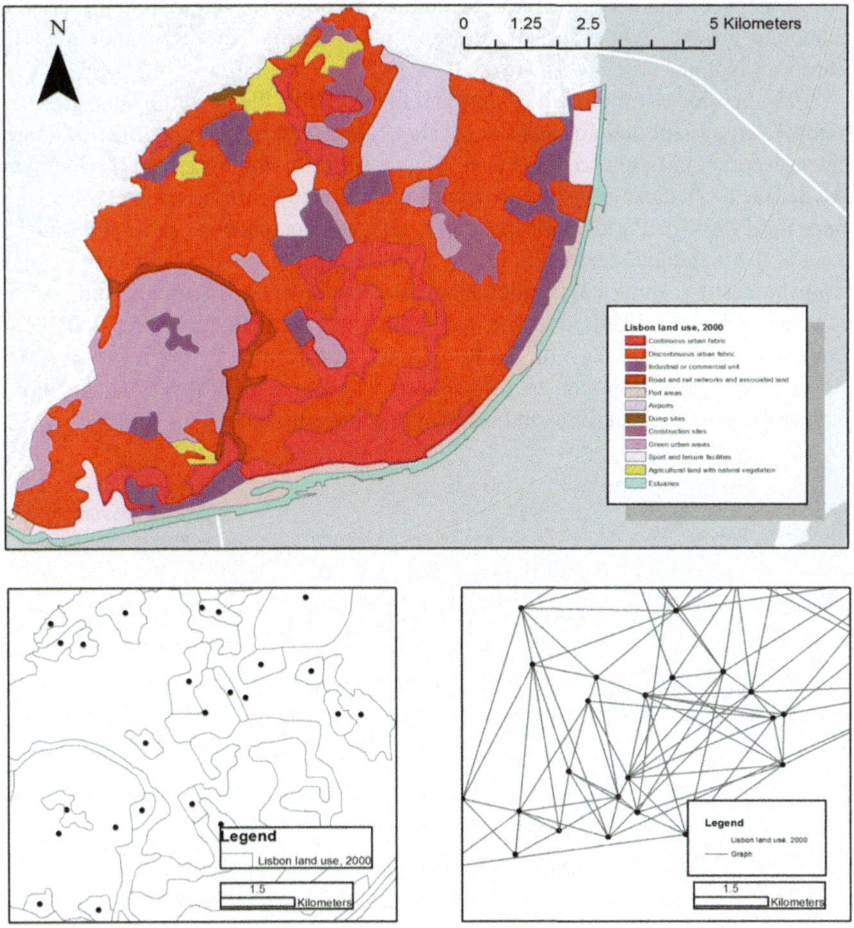

Fig. 6.7 Geographic abstraction of land use into graph

6.5 Further Modelling Results

Lisbon, the capital of Portugal, is one of the most vibrant and diverse urban regions of Portugal. Since the early 1980s the socioeconomic changes in the city have led to a fragmentation of some urban areas and increase of urban land in the city nexus. Portugal as a country has changed dramatically since the end of its dictatorship in 1974, and several important moments in the Portuguese political and economic scene have contributed to what Portugal is today at a land use level.

With the integration of Portugal in the European Union in 1984, land use has predominantly become urban land, and rural exodus has led to abandonment of agricultural and rural areas. In the mid-1980s, new urban regions were formed, and the tourist industry shaped most of the Portuguese coastal areas into what it is today. The strategy of the service sector accumulated allowed cities to become hubs of economic growth, while agricultural land was reformed through funding of the European Union.

After the successful growth of Portugal in the 1990s, the new millennia gave rise to protected agricultural areas and ecosystems, followed by a stronger legislation on the protection land use and land cover. The recent economic recession has led in the last decade to a stagnation of urban sprawl, and the analysis of urban land use change since the 1990s for Portugal clearly shows that this is a tendency generally found in most of the mainland, deemed to follow over the next years. The land use of Lisbon is an interesting empirical application, given its diversity of urban land use, its concentrated city nuclei, and the availability of datasets from CORINE Land Cover. A closer analysis on the resulting graph transformed by the Kamada-Kawai algorithm visually responds to the adjacency relations found over all the urban metropolis of Lisbon in 1990 and respective change in 2000 (Figs. 6.8 and 6.9).

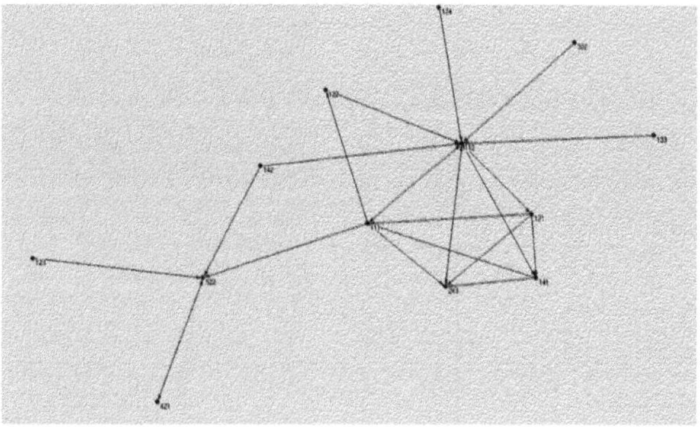

Fig. 6.8 Land use in Lisbon in 1990

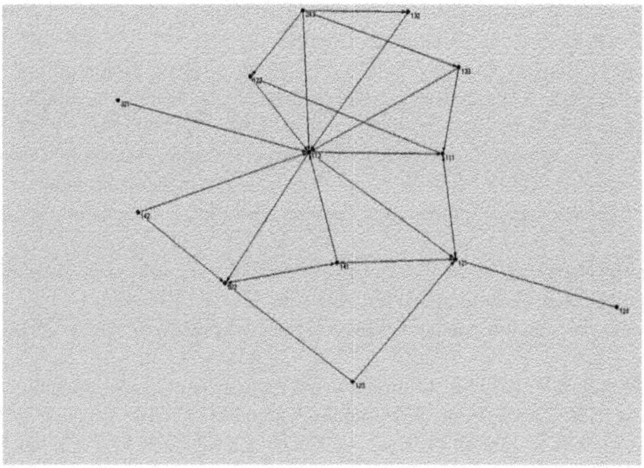

111 – Continuous urban fabric
112 – Discontinuous urban fabric
121 – Industrial or commercial units
122 – Road and rail networks
123 – Port areas
124 – Airports
133 – Construction sites
141 – Green urban areas
142 – Sport and leisure facilities
243 – Agriculture and natural vegetation
322 – Moors and heathland
421 – Salt marshes
522 – Estuaries

Fig. 6.9 Land use in Lisbon in 2000

We can see that a construct of interrelated land use types is formed between anthropogenic land use, corresponding to continuous urban fabric, industrial commercial units, and the attempt to add green spaces within the urban area. Agricultural land and natural vegetation are also considered in the context of existing urban areas. This integrated vision in the 1990s is strongly linked to the plans of the national inventory of agricultural land as explored in Vaz et al. (Vaz et al. 2012a, b). Sport and leisure facilities are fundamentally centered along the proximity of the coastal areas, and of natural environment, while forming quite an independent link. Discontinuous urban fabric seems to be of central importance in 1990 as well as in 2000.

This relation to discontinuous urban fabric fosters the concept that land use is in a permanent transition process, depending on the current economic, social, and environmental constraints within the urban nexus. The differences in the morphology of adjacency found in 2000 are evident. A much more centered role seems to exist, where land use is interchanged among different land use types and expressed by a land use mix, where all land use types are interlinked with exception of the leisure facilities as well as the appearance of a new class, relating to natural grassland. From 1990 to 2000, Lisbon land use types have become much more functional, expressing by the number of connections within the graph, leading to a more heterogeneous city combining availability of infrastructure, urban fabric, and the natural environment.

References

Arthurs D, Cassidy E, Davis C, Wolfe DA (2009) Indicators in support of innovation cluster policy. Int J Technol Manag 46(3–4):263–279

Batty M (2005) Cities and complexity: understanding cities with cellular automata, agent-based models, and fractals. MIT Press, Cambridge, MA

Cooke P, Braczyk HI, Heidenreich M (eds) (2004) Regional innovation systems: the role of governances in a globalized world, 2nd edn. Routledge, London

Cusimano M, Marshall S, Rinner C, Jiang D, Chipman M (2010) Patterns of urban violent injury: a spatio-temporal analysis. PLoS ONE 5(1):e8669. https://doi.org/10.1371/journal.pone.0008669

Davidson C (1998) Issues in measuring landscape fragmentation. Wildlife Soc Bull 26(1):32–37

Davis C (2008) Structural model of a regional information technology innovation cluster. In: Paper presented at the 25th Celebration DRUID Conference on Entrepreneurship and Innovation – Organizations, Institution, Systems and Region, Copenhagen, CBS, Denmark, June 17–20, 2008

de Groot R (2006) Function-analysis and valuation as a tool to assess land use conflicts in planning for sustainable, multi-functional landscapes. Landsc Urban Plann 75(3–4):175–186

Fisher M (2006) Innovation, networks, and knowledge spillovers: selected essays. Springer-Verlag, New York

Galindo PV, Noronha Vaz T, Nijkamp P (2011) Institutional capacity to dynamically innovate: an application to the Portuguese case. Technol Forecast Social Change 78(1):3–12

Han J, Fontanos P, Fukushi K, Herath S, Heeren N, Naso V, Cecchi C, Edwards P, Takeuchi K (2012) Innovation for sustainability: toward a sustainable urban future in industrialized cities. Sustain Sci 7:91–100

Jankowski P (1995) Integrating geographical information systems and multiple criteria decision-making methods. Int J Geogr Inf Syst 9(3):251–273

Litman T, Burwell D (2006) Issues in sustainable transportation. Int J Glob Enviro Issues 6(4):331–347

Moilanen A, Hanski I (2002) On the use of connectivity measures in spatial ecology. Oikos 95(1):147–151

Monteiro P, Noronha T, Neto P (2013) The idiosyncratic nature of maritime clusters: an essay towards their possible differentiation. Portuguese J Manag Stud 17(1):7–38. Lisboa

Nagendra H, Munroe D, Southword J (2004) From pattern to process: landscape fragmentation and the analysis of land use/land cover change, Agriculture. Ecosyst Environ 101(2–3):111–115

Nicholls W (2008) The urban question revisited: the importance of cities for social movements. Int J Urban Reg Res 32(4):841859

Noronha Vaz T, Galindo PV, Vaz E, Nijkamp P (2013) Innovative firms behind the regions: analysis of regional innovation performance in Portugal by external logistic biplots. Eur Urban Reg Stud 22(3):329–344

O'Neill R, Kahn JR (2000) Homo economus as a keystone species. BioScience 50(4):333–337

Roseland M (1997) Dimensions of the eco-city. Cities 14(4):197–202

Rosenthal SS, Strange W (2001) The determinants of agglomeration. J Urban Econ 50(2):191–229

Teigland R, Schenkel A (2006) Exploring the role of communities of practice in regional innovation systems. In: Coakes E, Clarke S (eds) The encyclopaedia of communities of practice in information and knowledge management. Idea Group, Hersley

Tobler W (1970) A computer movie simulating urban growth in the detroit region. Econ Geogr 46:234–240

Vaz E, Caetano M, Nijkamp P, Painho M (2012a) A multi-scenario prospection of urban change – a study on urban growth in the Algarve. Landsc Urban Plann 104(2):201–211

Vaz E, Noronha T, Nijkamp P (2012b) The use of gravity concepts for agricultural land-use dynamics: a case study on the Algarve. Int J Foresight Innov Policy 8(2-3):262–271

Xiao J, Shen Y, Ge J, Tateishi R, Tang C, Liang Y, Huang Z (2006) Evaluating urban expansion and land use change in Shijiazhuang, China, by using GIS and remote sensing. Landsc Urban Plann 75(1–2):69–80

Chapter 7
Southern European Coastal Environments: An Assessment of Portugal

Abstract Recent studies in urbanization have shown urban sprawl to be a contemporary phenomenon, not only tied to population increase, but rather a result of a complex changing landscape of cities in transition defining their economic patterns and their niches. In detriment to the coastal landscapes, urban sprawl has led to loss of natural and agricultural landscapes in southern Europe, particularly in peripheral regions such as the Algarve, where the recent economic recession has been a silent witness of loss of economic activity. To enhance the Algarve's sustainable development, this research assessed past urban growth on the basis of the MURBANDY project, covering urban footprint since 1972 for the Algarve. An urban growth model was implemented by means of assessing a new method for urban growth modeling based on percolation theory. The result is an assessment of multi-temporal change in the Algarve, assessing the urban growth rates for the Algarve's coastal stretches and the impact on its urban footprint. Using the MURBANDY stretch as a laboratory to agricultural and urban land use change, we analyze agricultural activity which has had little interest in a context of national economic development. We assess the changes in wetland systems and coastal change for the Portuguese mainland. The expanding importance of coastal zones, where over 80% of anthropogenic activity occurs, must be consolidated with the agricultural activity while not neglecting the importance of wetland and water systems in such areas. This chapter adopts a spatially explicit model to assess (i) coastal change in the Portuguese mainland, (ii) wetland loss in the coastal regions, (iii) depreciation of agriculture in detriment of urbanization processes; we further conclude on the importance of maintaining agricultural activity to support wetland sustainability, fomenting endogenous growth as a result avoiding unmanned urban planning. This integrative approach models the efficiency of wetland management for agricultural activity at regional level and can be applied to similar regions facing similar challenges.

Keywords Historical landscapes · Regional growth · Archaeological heritage · Southern Europe · Regional sustainability

E. Vaz, T. de Noronha, *Sustainable Development in Southern Europe*,
https://doi.org/10.1007/978-3-662-62177-6_7

7.1 Introduction

7.1.1 Land Use and Urbanization

Europe's increasingly changing urban perimeter has led to a growing debate on the sustainability of the rural-urban landscape given the role of cities and interactions of society and economy (Verhoef and Nijkamp 2004). In southern Europe, failure in urban planning has been a result of fragile economic growth, allied to the consequences of being peripheral regions (Minca 2003). The lack of consistent and long-term planning policies has further accrued to the deterioration of the coastal landscapes of southern Europe. Loss of agricultural land has been one of the main consequences of the creation of anthropogenic habitats in last decades (Vaz and Bowman 2013). The impact of agriculture has brought severe consequences on the stability of wetland systems. Wetland is of utmost importance for ecological sustainability and preservation of ecosystem landscapes, especially in regions where tourism activity is relevant for economic development (Arvela et al. 2018; Samora-Arvela et al. 2020). While agricultural land loss seems to be an ongoing trend at global level, it is up to decision-making to reverse these trends, by means of articulating spatial information into planning, as to induce a more efficient strategy to cope with environmental degradation.

Several landscape-scale and land use considerations should be taken into account when perceiving value to wetlands and the triangle of environment, economic growth, and wetlands. In this sense the best compromise for wetlands is found in the juncture of moderate anthropogenic activity, as to overburden wetland functions (Vaz et al. 2013). Wetland systems in Portugal are strongly linked to the history and culture of traditional agricultural activities (Vaz et al. 2015). While no registry of wetlands exists that were transformed into agricultural land use, a total of 130,943 ha are wetland systems.

In the Algarve, where a mass expansion of urban growth occurred during the 1980s and 1990s, a growing concern has risen since the economic crisis. As most of the existing infrastructure is becoming obsolete with the decreasing demand, existing infrastructure to support tourism in the coastal regions is becoming brownfields (Loures and Vaz 2018) permeating fragile land use and leading to coastal vulnerability. New models to measure the spatial and land use dynamics at coastal level are thus of utmost importance, to avoid the deleterious effects of excessive urbanization as well as understanding of impacts on the environment. A better understanding in coastal regions of the southern Europe and of Mediterranean sprawl should thus be interpreted in line with factors of dispersion and compactness while assessing the strengths and weaknesses of present urban regions (Salvati and Morelli 2014). A growing concern attributed by the socioeconomic construct on urbanized areas in coastal stretches has followed with the importance to understand better the urban and rural interactions in Europe's south (Aguilera et al. 2011), as well as to envision the challenges urban regions face in the future, and what may be expected given phenomena such as urban sprawl and land use change in the decades to come.

Portugal has reinforced its regional level by municipal plans since the 1980s, establishing a framework designated as the National Policy and Territorial Management Programme, addressing the challenges of urban sprawl in Portugal at regional level, taking as an example the umbrella of the European Union, and attempting to avoid unnecessary urban sprawl in protected and ecological sound areas. This has been achieved with some success, mitigating urban sprawl at NUTS III areas, although Portugal as a country has been the largest changing urban landscapes in Europe. Attempts in modelling the urban change in the Algarve have been carried out given the interest in understanding the regional dynamics of urban change in this region (Petrov et al. 2009; de Noronha Vaz et al. 2012). This calls out for new modelling approaches that allow to portray urban change based on novel techniques that equate not increasing urban growth, but, rather, locational urban sprawl. The aim of this chapter is to combine concepts found in theoretical physics and advanced statistics to generate a Percolation model for urban sprawl in the Algarve. The choice of the Algarve is linked to the current stagnation of urban growth that traditional assemblies of urban growth models have not been able to efficiently acquire. The integrated percolation framework proposed in this chapter further allows a rapid assessment on urban growth in regions where data availability is scarce and where urban sprawl is an imminent consequence but bound to lead to stagnation and generation of obsolete infrastructures. The MURBANDY datasets will be used in this study to generate an integrative urban growth model, leading to a better understanding of the consequences of the coastal fringe of the Algarve and impacts of urban sprawl in this region. The spatially explicit characteristics of rule-based cellular automata (CA) have widely been used in a vast array of different applications (Chaudhuri 1997) in particular when coupling forecasting capabilities (Herold et al. 2003). Given the results of stochastic outcomes, these models allow a geographical understanding of the probabilistic outcomes of urban areas (Mundi and Murayama 2010; de Noronha Vaz et al. 2012; Onilude and Vaz 2020), enabling accurate outcomes for planning and landscape purposes and better representing the complex interaction of urban regions (Batty and Xie 1994). This is a result of their bottom-up approach, where the state of a cell iterates in time, allowing to construct scenarios for urban change (de Noronha Vaz et al. 2012). Such an approach to land use is of paramount importance given the consequences of urban sprawl on health (Griffin et al. 2013; Vaz et al. 2020a, b) and natural environment where cellular automata allow modelling different phenomena at different scales (Engelen et al. 1995). In countries such as Portugal: (i) Rural transitions have led to disperse population; (ii) the increase of lack of infrastructural connectivity and (iii) new urban developments to respond to the tourism industry have been identified in suburban regions. In a European framework, this fortifies further the importance of spatial modelling approaches to identify the consequences on land use for better decision-making (Koomen et al. 2007). In the case of the Algarve, consequences lead to the importance of a multilevel assessment of the factors behind urban growth and justify the application of land use models developed in recent years that greatly contribute to better planning and management of land use (Vaz and Arsanjani 2015; Vaz and Jokar Arsanjani 2015). Nonlinear approaches to model urban sprawl and growth relating spatial change (Arsanjani et al. 2011) have been in the forefront of sharing useful information for planning purposes, enabling better decision-making in line with the results of urban complexity (Batty 2007).

7.1.2 The Economic Context

In fact, the application of the R&D policy brought to Europe the preference for those regions able to produce high-tech industries, clearly located in the urban, already more developed areas. Furthermore, frequently, those public policies for regional development created to increase competitiveness of less favored regions crossed targets at national level. Technological modernization occurring in urban areas transmitted spillover effects that reduced the eventual difficult to get positive impacts of other regional policies – the case of PRIME, a specific program for the modernization of Portuguese enterprises, illustrates this situation (REF).

The preceding debate marked much of the effective regional development of many European regions. After more than 20 years of application of a complete package scheme of European supports, Portugal is an excellent example to show how public policies in Europe have rather served to augment regional and by sector disparities: Litoral versus interior and urban concentration versus rural diversification (REF).

Portugal is small yet a diverse country, with weak growth and limited public spending capacity. It is marked by a long tradition of centralized governance and no elected regional level. The country population is of 10.6 million people (2006) from which 50% is living in PUR regions, 24% is living in IR, and 26% is living in PRR. Portugal is divided in several distinct regions and two autonomous regions (Azores and Madeira). The mainland is constituted by the North, Central, Lisbon, Alentejo, and Algarve regions.

It should also be added that the local level in Portugal has little to do with the environment of regional policy: a multitude of 308 municipalities, with a municipal assembly, a mayor, and an executive council elected every 4 years. Their main responsibilities are related to the management of collective equipment and basic infrastructure.

Regional disparities in Portugal are represented by a vertical dichotomy between a dense and dynamic urban coast and a desert and declining rural interior. Between 1995 and 2006, population density increased markedly in urban regions and in the intermediate regions located next to the urban regions (Fig. 7.1).

In the Portuguese GDP per capita, regional disparities seem linked to the economic cycle: During years of robust economic growth (1995–2000), the regional dispersion increased; when the economy slowed down, regional disparities also decreased. Due to the large contribution of Lisbon to national output, regional disparities and national growth rates are both highly sensitive to Lisbon's economic performance (Fig. 7.2).

The disparities found in Portugal are quite significant in regard to the gross domestic product. Further assessment of the Gini index shows a disparity of 0.57

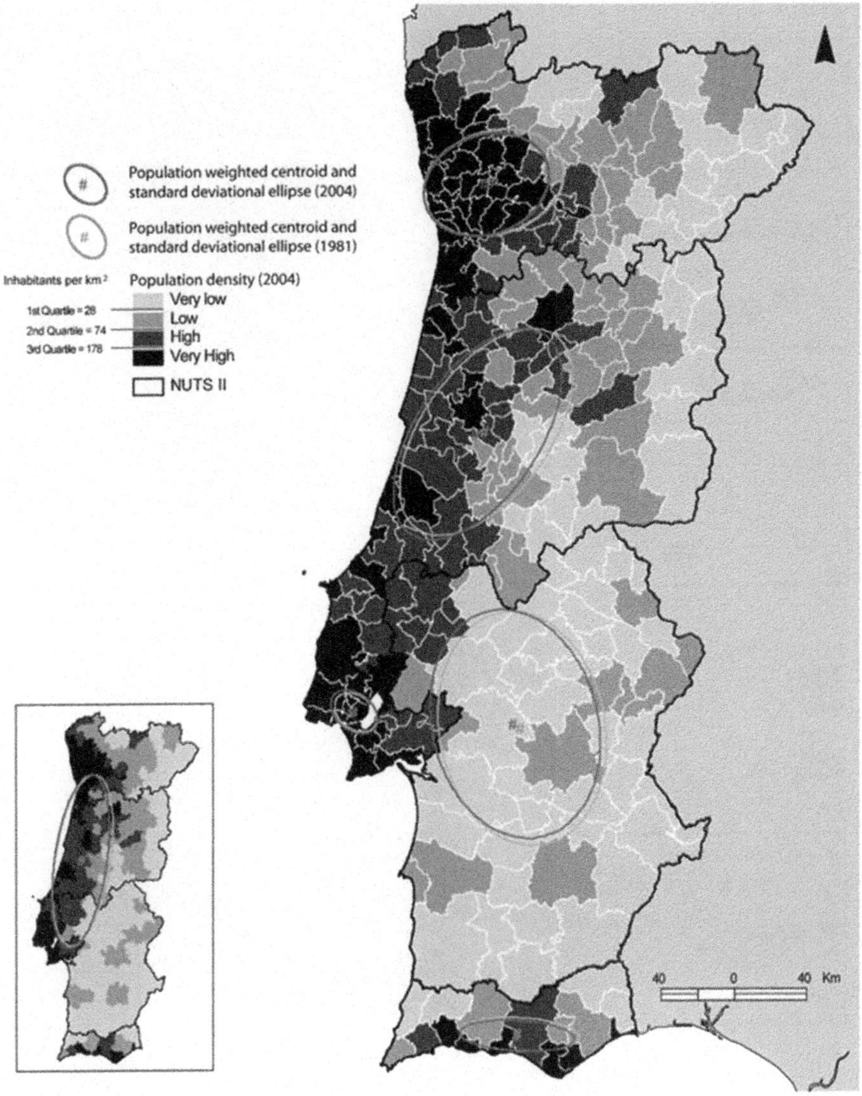

Fig. 7.1 Population density in Portugal, 2004
Source: OECD Territorial Reviews, 2008

with a difference of 0.09 from the OECD. This is fueled by the fact that the two of the Portuguese metropolitan regions of Lisbon and Oporto generate 43% of the GDP of the country.

Regional disparities in GDP are in turn closely linked with the pattern of regional specialization. Not surprisingly, Portuguese urban regions devote a higher share of their total employment to service activities than rural and intermediate regions.

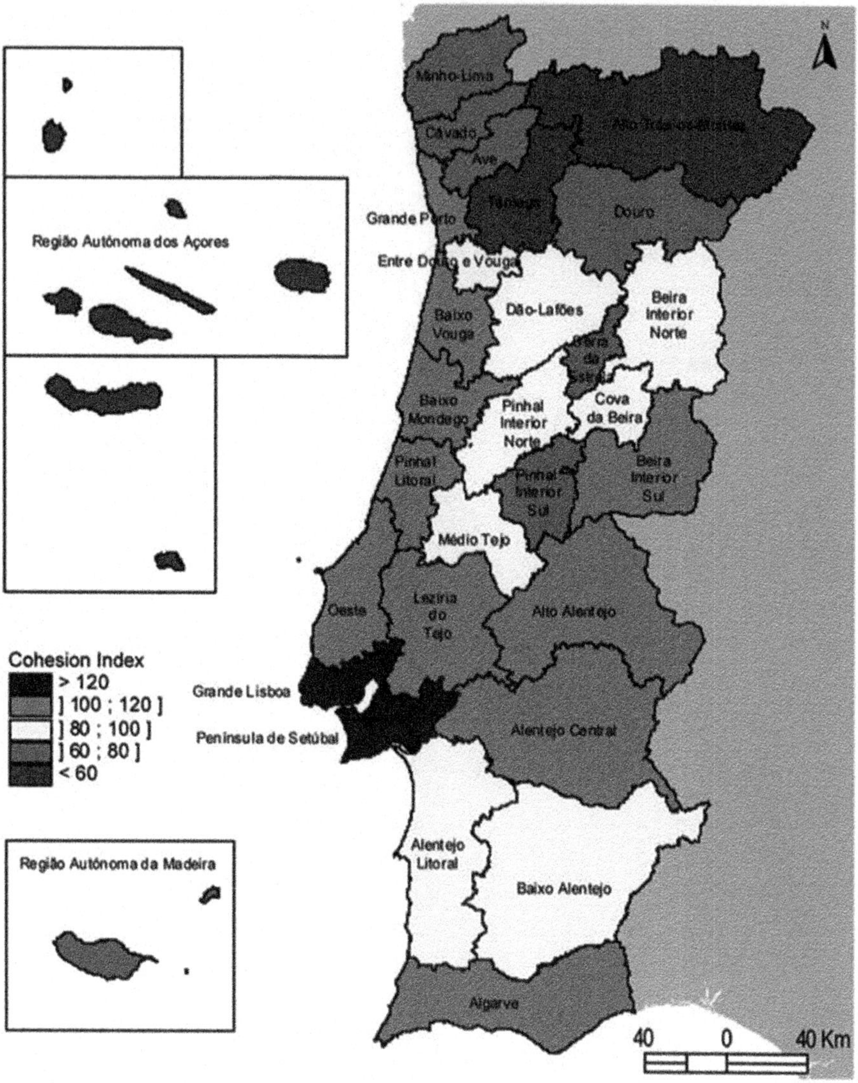

Fig. 7.2 Map of Portuguese regions classified according to the "cohesion" index
Source: OECD Territorial Reviews, 2008

7.2 Study Area

The region of the Algarve is the most southern region of Portugal and a region of pristine natural landscapes and beauty, encompassing a unique history and biodiversity that has led to a large-scale tourism industry back in the 1960s. With its district capital of Faro, located at 37°0′52″N 7°56′7″W, the average elevation is of

11m, sharing its strong coastal traditions with the active fishing industries dating back to the nineteenth century. The Algarve offers one of the highest numbers of protected regions by the NATURA network in Europe, with a total of 38.6% of the Algarve territory part of this natural reserve. This unique ecological diversity has led to the continued importance of defining a sustainable tourism industry, where the transformation of economy, society, and environment must be well balanced with rural activities and the correct planning of coastal regions. Urban areas, however, have become a great pressure on this natural environment, boosted by a population increase between 1960 and 1991 of almost 60% in the urban cores of the Algarve. The current economic recession has led to severe abandonment of economic activity linked to tourism and residues of obsolete touristic infrastructures that have to find a new use, before transforming into brownfields in the vicinities of natural protected landscapes.

7.2.1 Data

The study has been performed for the Portuguese coastline to show how it has changed over time based on satellite imagery that is available on the Global Land Cover Facility (GLCF-www.glcf.umiacs.umd.edu).

The remarkable effort conducted by NASA in 1998 to create the Global Land Cover Facility has led to an outstanding inventory of remote sensing imagery free of charge. This has brought a plethora of different land use analysis studies that intertwine different scientific communities to understand not only changes along the Earth's surface, but an attempt to mitigate regional issues of existing asymmetries at regional level pertaining to inefficient distribution of urban land and other environmentally relevant land use/land cover categories. The Global Land Cover Facility integrates several distinct sources into their database:

- Satellite imagery (Aster, Ikonos, Quickbird, Orbview, Landsat, Modis, SRTM)
- Products derived from satellite imagery such as Forest Change Products, Coastal Marsh Health Index, Flood Maps, Vegetative Cover Conversion, Vegetation Continuous Fields, Vegetation Indices
- Vector products

The main advantage of the Global Land Cover Facility is that the user is allowed to search and select free of charge data based upon date, location, or a mixture of parameters and then download via web application called Earth Science Data Interface (ESDI) without needing to register.

Data that has been used to extract shoreline in multi-temporal (1988/1989, 2000, 2007) are Landsat 5 and 7 satellite imagery selected and downloaded from the Global Land Cover Facility sources. In order to cover all required area, one image acquired in 1988 needed to be added to dataset of 1989 (Fig. 7.3, Table 7.1).

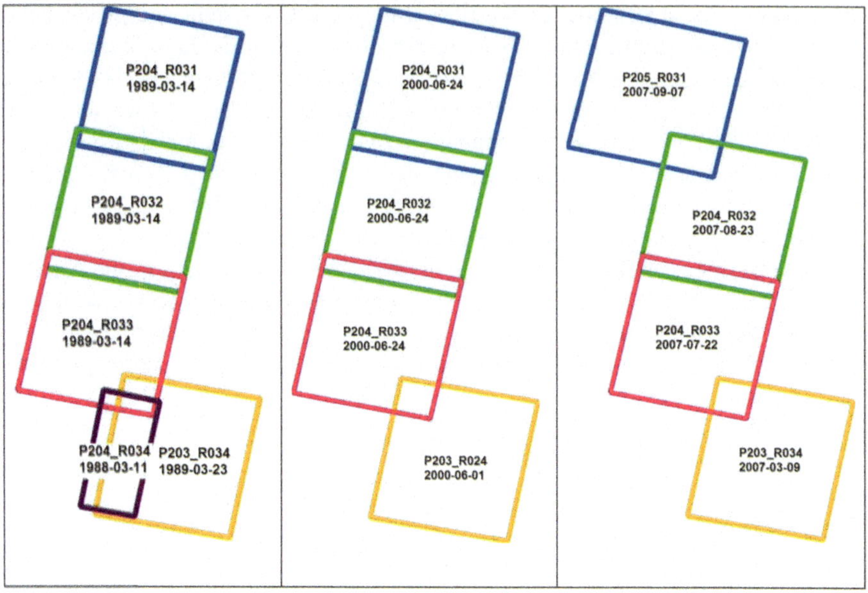

Fig. 7.3 Overview of the Landsat sample images that has been used

Table 7.1 Basic metadata of the Landsat sample images

	Satellite (sensor)	Acquisition date	Path/row	Resolution [m]	Map projection (ellipsoid, datum)
1988/ 1989	Landsat 5 (TM)	1989-03-14	204/031	30	UTM, WGS 84, ZONE 29 (WGS 84)
		1989-03-14	204/032		
		1989-03-14	204/033		
		1988-03-11	204/034		
		1989-03-23	203/034		
2000	Landsat 7 (ETM+)	2000-06-24	204/031	28.5	
		2000-06-24	204/032		
		2000-06-24	204/033		
		2000-06-01	203/034		
2007	Landsat 7 (ETM+)	2007-09-07	204/031	28.5	
	Landsat 5 (TM)	2007-08-23	204/032	30	
		2007-07-22	204/033		
		2007-03-09	203/034		

7.2.2 *Methodology*

Several methods have been developed for the coastline detection using satellite imagery. The simplest one is based on one of infrared bands where the reflectance of the water is nearly equal to zero and the reflectance of absolute majority of land coverage is greater than water. It can be seen on the histogram as a two peaks, exhibiting a strong contrast between land and water (Fig. 7.4).

The transition zone is the effect of mixed pixels and moisture regimes between land and water that's why it is quite difficult to find a proper threshold value. To reduce influence of the threshold value choice and to improve extraction of the coastline, the second condition is introduced: the ratio b2/b5 is greater than one for water and less than one for land in large areas of coastal zone.

As a result of those two steps, two binary images are obtained with the land pixel equal to 0 and the water pixel equal to 1. By multiplying those two images, one is obtained that represents the coastline accurately. In the last step, export from binary to vector map is required as it is shown in Fig. 7.5.

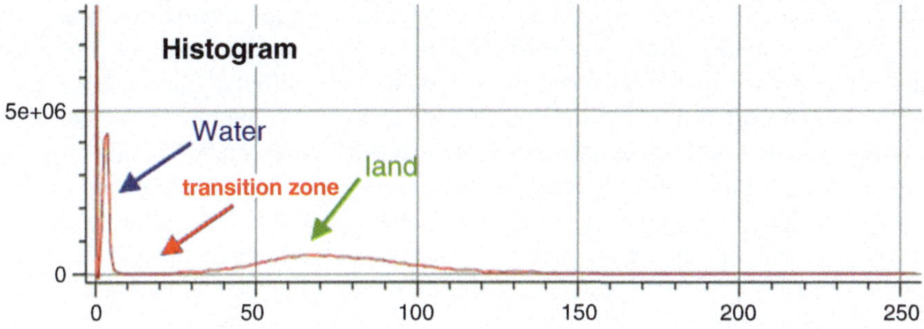

Fig. 7.4 Histogram of the band 5, Landsat TM-1989

Fig. 7.5 Methodology of extracting coastlines from images

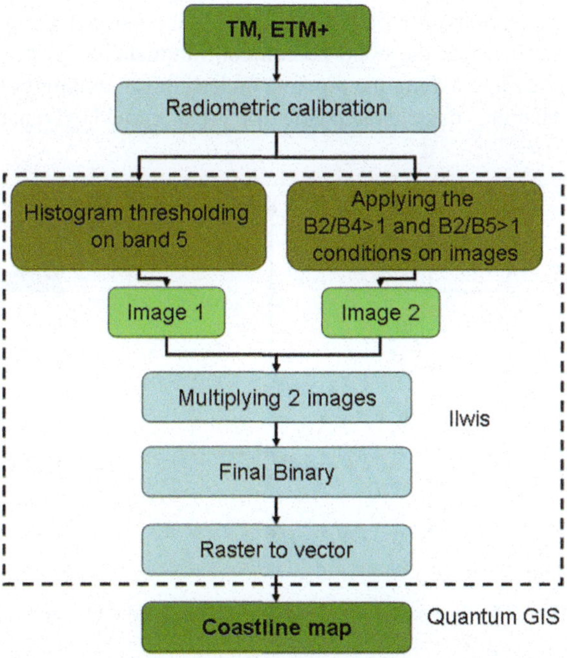

7.3 Methodology

7.3.1 *Integrating Percolation*

The changing morphology of cities poses an interesting subject to be examined in line with percolation theory and statistical physics (Makse et al. 1998). One of the main similarities that make percolation to be spatially explicit is its characteristic of connectivity. This connectivity assumes the same characteristics as urbanization processes, in which the concept of distance decay applies similarly to the porosity of sediment, where less porous materials are equally a representation of regions less prone to become urbanized. While traditionally percolation represents the classical example of infiltration towards porosity, and has thus become an important instrument for material science, it is the universality of the concept of connectivity and the degree of infiltration given properties of porosity (Leuenberger and Leu 1992) that deems to be one of the universal properties in many different approaches where clusters and orders of clustering prevail for urban areas. Figure 7.6 shows this intrinsic relation between lattices and clusters that are representation of urban sprawl phenomena. While, to the left, bond percolation is represented by lattices and vectors that show their properties of connectivity, the figure to the right is a similar representation but transformed into a cellular model, traditionally adapted in urban growth models, representing an abstraction of urban sprawl agglomerates.

This allows not only visualizing the agglomeration of urban sprawl, but better understanding the interaction of urban clusters given a set of determinants that represent the concept of porosity. This concept is translated into a Markov transition chain, as expanded in the methodology section, and enables an approximation of a given probability to an adjacent cell that has a given probability of becoming urban, in line with the bond percolation rationale, of becoming more or less porous. It is therefore a dynamic process, underlying an assumption of spatial explicit dynamics, enabling the characteristics of urban sprawl, given the laws of a certain type of

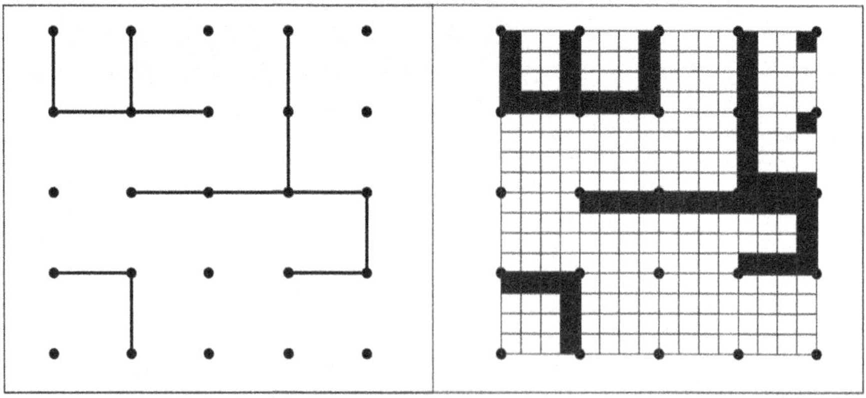

Fig. 7.6 Representation of bond percolation and urban cells

percolation, such as bond percolation, as the transformation of urban change affects the density approximation of the adjacent cell to previously existent urban areas. This is very similar to the natural phenomenon of percolation and is applied in this chapter to one of the most urban dynamic regions in the last decades, the Algarve. The chapter starts out by explaining the land use context of the Algarve region, applying then a Markov transition matrix and generating a model of bond percolation at regional level. Based on findings, it becomes possible to understand which segments are more plausible to change in the coming decades, bringing new insights to micro-spatial analysis of urban processes.

Several components construct the urban growth model to forecast urban change in the Algarve (Fig. 7.7). The workflow starts with the composition of several land use classified datasets from the MURBANDY project, leading to (i) definition of land use change and creation of Markov transition maps, (ii) the order of change as to

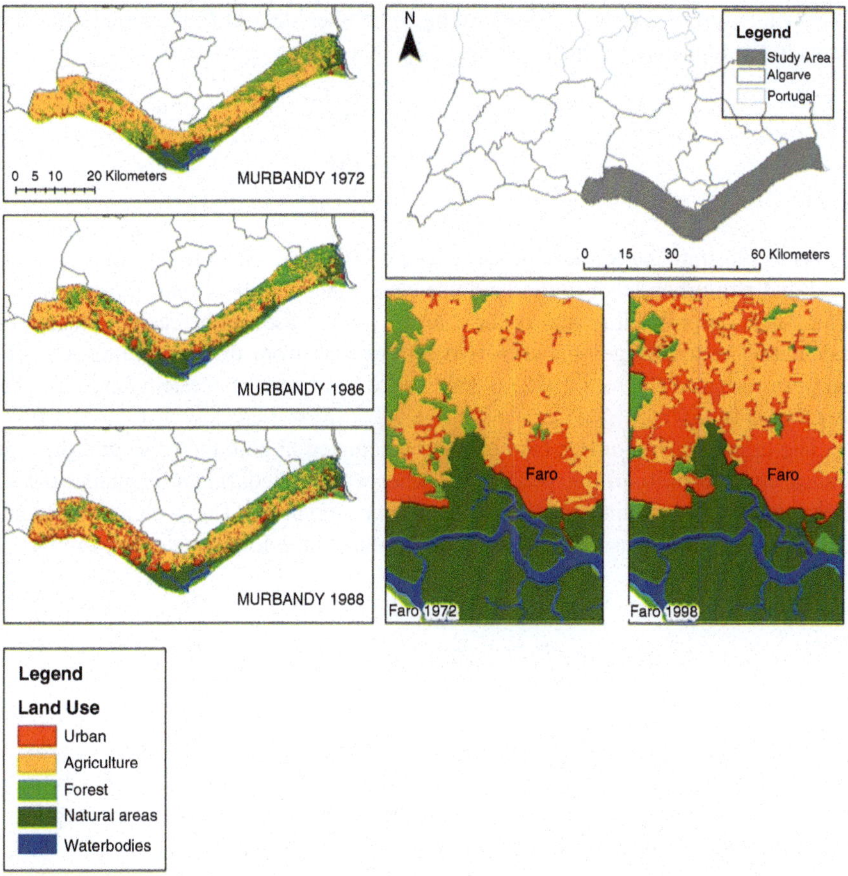

Fig. 7.7 MURBANDY land use in 1972, 986, 1998 (left) zoom of land use for the district capital of Faro (below right)

define the percolation model, and (iii) forecast urban growth up until 2020. Step ii) presupposes the validation through calculating a kappa statistic. This statistic allowed to validate the accuracy of the predicted bond percolation model.

In Markov processes, the transition state of a target at time t is only relative to its previous state at time $t - 1$, but not relative to the state at time $t - 2$, $t - n$. The Markov process can be applied to time series as well as spatial series. In a Markov chain, the probability of the transition from a state at time t to another state at time t +1 is called the state transition probability. If the occurrence of an event E_j depends on the condition of another event E_i, its transition probability $P(E_i \rightarrow E_j)$ is the conditional probability $P(E_i/E_j)$, leading to:

$$P(E_i \rightarrow E_j) = P\left(\frac{E_i}{E_j}\right) = P_{ij} \tag{7.1}$$

To obtain the transition probability matrix, the transition probability P_{ij} where $(i, k = 1, 2, \ldots, n)$ of a state E_i transforming to another state in E_j is carried out. This leads to the transition probability matrix P_{ij}, that can be represented as:

$$P = \begin{matrix} p_{11} & \cdots & p_{1N} \\ p_{11} & \cdots & p_{2N} \\ p_{11} & \cdots & p_{3N} \end{matrix} \tag{7.2}$$

As to illustrate the Markov process, let's consider two land use states, the urban land and non-urban land, in the land use classes denoted by 1 and 0, respectively. The transition probability matrix for changes in land use can be easily obtained by calculating state changes between two sets of data from different years. For this purpose, urban data from 1972 to 1986 were used for the western region of the Algarve (Fig. 7.7).

The urban and non-urban land use was calculated in both the years and then the transition probability of one state to the other was applied; this originated a probability matrix, integrating a weighted distance approach. The weighted distance approach with the transition probabilities originated the following matrix m:

$$m = \begin{matrix} 0.7071 & 1 & 0.7071 \\ 1 & 1 & 1 \\ 0.7071 & 1 & 0.7071 \end{matrix} \tag{7.3}$$

In addition to the Markov transition probabilities, a weighted distance approach was used. This approach assumes that the influence of a neighbor cell depends on its distance to the target cell. Thus, the weighted index is used to define the degree of influences (Xiaoying et al. 2004).

The neighborhood of a cell (surrounding cells) influences the transition of this cell into other classes in the next time step. The cells located further away have a smaller effect than cells closer to the center cell. The transition rules are the core of the CA and determine if, and how, the state of each cell in the next time step changes:

$$Nwj = \sum_{n=1}^{m} \frac{1}{d_n} \tag{7.4}$$

Nwj: The total weight for the state j
m: The number of the state j in a defined neighborhood
n d: The distance between the nth state j and the center cell

From the above, the transition index for a cell of the change in land use can be expressed as:

$$CI = Max(P_{ii}Nwj) \tag{7.5}$$

$$\mathrm{J} = 1, \quad 2, \quad 3 \cdots \mathrm{n}$$

Using the land use state matrix, a distance matrix is obtained, and the weighted distance of every state is determined. This is combined with the transition probability to get the transition index of each cell.

7.3.2 *Percolation Probability Model*

This model is based on bond percolation method (Berkowitz and Ewing 1998). Percolation theory deals with the numbers and properties of the clusters formed when sites are occupied with probability p.

We consider the d-dimensional lattice L^d=(Z^d, E^d) where the set of edges E^d connects sites (x, y) = ((x1, . . ., xd), (y1, . . ., yd)) located at the vertices of Z^d for which the distance, defined by (Grimmett 1999):

$$\delta(x, y) = \sum_{i=1}^{d} | x_i - y_i | \tag{7.6}$$

is no more than one: _(x, y) $\leq$ 1. The edges of E^d connect thus adjacent vertices of Z^d. Let 0 $\leq$ p $\leq$ 1. We declare an edge of E^d to be open with probability p and closed otherwise probability 1-p, independently of all other edges.

We denote by C(x) the part of L^d containing the set of vertices connected by open paths to vertex x and the open edges of E^d connecting such vertices. By translation invariance of the lattice and the probability measure P^p, the distribution of C(x) does not depend on the vertex x. We therefore take in general x = 0 and denote by C the open cluster at the origin: C = C(0).

If A and B are sets of vertices of Ld, we write A $\leftrightarrow$ B to express the fact that there exists an open path connecting some vertex of A to some vertex of B. For example, C (x) = {$y \in$ Zd| x $\leftrightarrow$ y}.

We write ∂A to denote the surface of A, which is the set of vertices of A which are adjacent to some vertex that does not belong to A. A typical subset of vertices is a box, defined as:

$$B(n) = [-n, n]^d = \left\{x \in \mathbb{Z}^d | \max_{1 \leq i \leq d} \{|x_i|\} \leq n\right\} \tag{7.7}$$

for some n $\in$ N $*$ = N\{0}. We write B(n, x) for the box x + B(n) having side length 2n and center at x. We will also work often with "diamond" boxes:

$$S(n) = \{x \in \mathbb{Z}^d | \delta(0, x) \leq n\} \tag{7.8}$$

or more general rectangular boxes. We also write S(n, x) for the diamond box x+S (n) centered in x.

The main quantity of interest in percolation theory is the probability that the origin belongs to a cluster with an infinite number of vertices:

$$\theta(p) = \mathbb{P}_p(|C| = \infty). \tag{7.9}$$

By space invariance, θ (p) is the probability that any node belongs to an infinite cluster:

$$p_c = \sup\{p | \theta(p) = 0\}. \tag{7.10}$$

The percolation threshold in L^2 is pc = ½.

This is the mathematical basis of percolation, where given a two-dimensional lattice of open and closed spaces, there are clusters, which are a group of nearest neighboring occupied sites. Percolation theory deals with the numbers and properties of the clusters formed when sites are occupied with probability p (Fig. 7.8).

There are four clusters in the above pic, one of seven occupies sites, one of three and two of one each.

In a land use model, sites occupied by the same land use class are being considered a cluster. All sites not occupied by a land use class are considered open sites which have a probability to be converted into that land use class.

Before application of the model, the clusters are identified, and for each cluster a bounding box is determined. The boundary cells are checked for their transition probabilities to determine the percolation from the cluster. This is repeated till the probability is lower than critical percolation probability, p_c. Once the critical probability is below 1/2, there is no more percolation, and the site remains open and doesn't join the cluster.

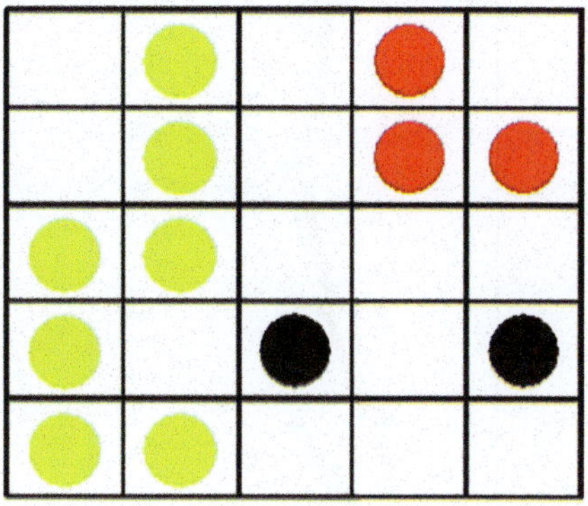

Fig. 7.8 A two-dimensional lattice showing different clusters

7.3.3 Coastal Erosion Impacts

The methodology discussed above was used for the purpose of our study. Shorelines from the year 1988/1989, 2000, and 2007 have been detected and extracted using two open source softwares. As an output of the first part (Fig. 7.8) performed in ILWIS (version 3.0), the vector form of the coastline was received. The results were then exported, as a shape file to Quantum GIS (1.7.2 version) for smoothing and improving the edge/boundary between water and land based on a color composite of 543 images which present a decent water-land interface. To evaluate the accuracy of this approach, usually comparison via visual interpretation to the ground truth map is done. In our case color composite, 543 map was a kind of a basis for accuracy evaluation.

The area of the study is extensive Portuguese coastline located on the Iberian Peninsula with the length of 1,793 km. The coastline is very well-known for steep and rough cliffs and friendly sandy beaches which make the cost very sensitive for the changes. Although acceleration or erosion rates depend on many natural factors such as weather, geology, and climate changes, human factor is also very important. Comparison of coastline erosion occurrence and the population density at the Portuguese coastline that is present on Fig. 7.9 shows a strong colligation of the erosion occurrence to the numbers of inhabitants.

7.4 Conclusions

7.4.1 The New Rurality in Southern Europe

A significant complexity of policies in the rural world is required to tackle systemically the challenges modern rural regions face in southern Europe. While the process of knowledge creation is an intrinsic part for sustainable development in the future, if spatial phenomena are not diligently monitored, impacts are

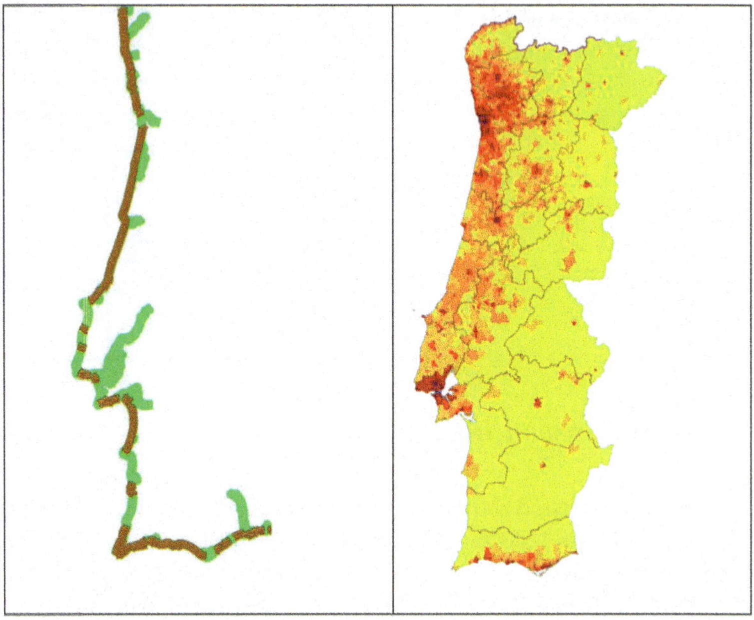

Fig. 7.9 The left part of the figure shows propensity to erosion, extracted from the satellite imagery acquired in 2007. The erosion was marked based on the coastlines extracted from multi-temporal satellite imagery (1989, 2000, 2007). To the right, the most populated regions of Portugal are shown

irreversible. The usage of comparative case studies, therefore, within the southern European framework is paramount to offer theoretical findings that may have relevant public policy implications, so that novel pathways towards sustainable development may be considered for immediate action. Urban sprawl and the change of coastal environments are relevant aspects for this understanding in lagging regions of southern Europe. Andalusia, for instance, a traditionally active rural region of Europe's south, has shown remarkable advances through its development and encouraged a significant amount of sustainable development by its integration of heritage, local culture, agriculture, and business innovation. A closer inspection of socioeconomic drivers (2003) shows that Andalusia, encompassing almost 90,000 km^2 and a population of 7.6 million, has had 11.3 percent of employed in the primary sector, followed by 10.8 percent in industry, 14.4 percent in the construction, and over 63.4 percent in services. The contribution of the primary sector to the regional product is about 6 percent in a context of 99.93 percent of small- and medium-sized firms. There are about 304 large corporations in the region. About 19.8 percent of the population has tertiary education, and as a result of public and business expenditures in R&D, the region has begun to significantly increase its levels of patent applications as a contribution to regional innovation.

Andalusia's agri-food industry is one of the bases for its successful regional development. Olives and its derivatives, fruits, wine, and vegetables utilize the environmental and climatic conditions at local level to gear a unique product of endogenous growth, resulting from the quality of its soils, historical traditions of production, as well as intensive and oftentimes innovative productive units that gear towards the exportation at national and international level. The agri-food industry represents as such a cornerstone of all industrial activities of the region, leveraging additionally from biotechnological innovation where bioprocesses and bioreactors become novel sectors of activity for research and development. Almost half of these new sectors of economic activity are allied to the agri-food industry in Andalusia. The European Union-funded Regional Innovation Technology Transfer System (RITTS – between 1994 and 1996) had a crucial role in the exploration of these sectors and enabled the design and strategy towards a synergy of the regional development agency (IFA) together with businesses, researchers, organizations, and governmental institutions to generate a blossoming flow for bio-innovation. This has generated at present a hub for scientific and technological advancement, where 10 universities, 22 technological centers, and 16 affiliated laboratories work with a total of seventy knowledge creation units for regional prosperity. CESEAND (Centro de Enlace para la Innovación del Sur de Europa) has monitored, mapped, and assessed these two decades of economic dynamics efficiently.

7.4.2 *Southern Europe's Coastal Spillovers*

The lesson learned with Andalusia's success is strongly linked with the capacity to innovate while respecting the historical path tendency of knowledge spillovers due to efficient investment strategies geared decades earlier towards sustainable development. This chapter has shown how the paradigm of understanding novel approaches to urbanization, as well as the possible effects of coastal change, has been enabled in a functional and sustainable design of policies. Within the changing characteristics of southern Europe's coastal areas, it is crucial to understand the original historical path tendencies of economic success, to incorporate these in a coherent policy structure. Urban sprawl, if not managed accordingly, behaves as a percolation model where the impacts on fragile and relevant land use may bring scarcity. Through the example of Andalusia's growth, we have clearly seen that innovation and development of enriching sectors is important, and knowledge spillovers in the rest of the south must be considered within the changing coastal environments.

7.4.3 *Reinventing the Rural South: Matching Urbanization with Climate Change*

A new integrated vision is allowing us to better manage and calibrate the rural world of tomorrow. The combination of geography, economics, spatial modelling, and the intention to monitor policies and impacts of these at local level given the advances in regional quantitative methods may hold new opportunities for local prosperity (Vaz and Agapito 2020). The geographical spillover is not evident: Project implementation and a coherent vision for decades may lead to a wider societal debate of the possibility for actors and scientists alike to collaborate with stakeholders for a new emergent relation in rural southern Europe. In the face of climate change (Vaz et al. 2013; Vaz and Bowman 2013), and the unavoidability of urbanization, it is important to consider the regional impacts of both urban growth (by using innovative ways of measuring urban sprawl) and the impacts of coastal erosion as to guarantee industries in the coming decades.

References

Aguilera F, Valenzuela LM, Botequilha-Leitão A (2011) Landscape metrics in the analysis of urban land use patterns: a case study in a Spanish metropolitan area. Landscape and Urban Planning 99:226–238

Arsanjani J, Kainz W, Mousivand A (2011) Tracking dynamic land use change using spatially explicit Markov Chain based on cellular automata: the case of Tehran. Int J Image Data Fusion 2:329–345

Arvela AFS, Ferreira JR, Oliveira R, Panagopoulos T, Vaz E (2018) Preferências de turistas por recreio e lazer na região do Algarve num contexto climático em mudança. Livro de atas do XVI Colóquio Ibérico de Geografia:766–774

Batty M (2007) Cities and complexity: understanding cities with cellular automata, agent-based models, and fractals. New York, The MIT Press

Batty M, Xie Y (1994) From cells to cities. Environ Plann B-Plann Des 21:531–538

Berkowitz B, Ewing RP (1998) Percolation theory and network modeling applications in soil physics. Surveys Geophys 19(1):23–72

Chaudhuri P (1997) Additive celular automata – theory and applications, volume 1. New York, John Wiley and Sons

de Noronha Vaz E, Nijkamp P, Painho M, Caetano M (2012) A multi-scenario forecast of urban change: a study on urban growth in the Algarve. Landsc Urban Plan 104(2):201–211

Engelen G, White R, Uljee I, Drazan P (1995) Using cellular automata for integrated modelling of socio-environmental systems. Environ Monitor Assess 34(2):203–214

Griffin BA, Eibner C, Bird CE, Jewell A, Margolis K, Shih R, Slaughter ME, Whitsel EA, Allison M, Escarce JJ (2013) The relationship between urban sprawl and coronary heart disease in women. Health Place 20:51–61

Grimmett GR (1999) Percolation, 2nd edn. Springer, Berlin

Herold M, Goldstein N, Clarke K (2003) The spatio-temporal form of urban growth: measurement, analysis and modeling. Remote Sens Environ 85:95–105

Koomen E, Stillwell J, Bakema A, Scholten HJ (2007) Modelling land-use change: progress and applications. Springer, Dordrecht

Leuenberger H, Leu R (1992) Formation of a tablet: a site and bond percolation phenomenon. J Pharma Sci 81(10):976–982

Loures L, Vaz E (2018) Exploring expert perception towards brownfield redevelopment benefits according to their typology. Habitat Int 72:66–76

Makse H, Andrade JS Jr, Batty M, Havlin H, Stanley E (1998) Modeling urban growth patterns with correlated percolation. Phys Rev E 58(6):7054–7062

Minca C (2003) Critical peripheries. Environ Plann D Soc Space 21(2):160–168

Mundia CN, Murayama Y (2010) Modeling spatial processes of urban growth in African cities: a case study of Nairobi city. Urban Geogr 31(2):259–272

Onilude O, Vaz E (2020) Data analysis of land use change and urban and rural impacts in Lagos state, Nigeria. Data 5(3):72

Petrov et al (2009) Urban land use scenarios for a tourist region in Europe: applying the MOLAND model to Algarve. Portugal Landsc Urban Plann 92:10–23

Salvatti L, Morelli V (2014) Unveiling urban sprawl in the Mediterranean region: towards a latent urban transformation? Int J Urban Reg Res 38(6):1935–1953

Samora-Arvela A, Ferreira J, Vaz E, Panagopoulos T (2020) Modeling nature-based and cultural recreation preferences in Mediterranean regions as opportunities for smart tourism and diversification. Sustainability 12(1):433

Vaz E, Arsanjani JJ (2015) Predicting urban growth of the greater Toronto area-coupling a markov cellular automata with document meta-analysis. J Environ Inf 25(2):71–80

Vaz E, Jokar Arsanjani J (2015) Crowdsourced mapping of land use in urban dense environments: an assessment of Toronto. The Canadian Geographer/Le Géographe canadien 59(2):246–255

Vaz E, Agapito D (2020) Recovering ancient landscapes in coastal zones for cultural tourism: a spatial analysis. In: Regional Intelligence. Springer, Cham, pp 9–28

Vaz E, Bowman L (2013) An application for regional coastal erosion processes in urban areas: a case study of the Golden horseshoe in Canada. Land 2(4):595–608

Vaz E, Walczynska A, Nijkamp P (2013) Regional challenges in tourist wetland systems: an integrated approach to the Ria Formosa in the Algarve, Portugal. Reg Environ Chang 13(1): 33–42

Vaz E, Painho M, Nijkamp P (2015) Linking agricultural policies with decision-making: a spatial approach. EuropeanPlanning Studies 23(4):733–745

Vaz E, Shaker RR, Cusimano MD, Loures L, Arsanjani JJ (2020a) Does land use and landscape contribute to selfharm? A sustainability cities framework. Data 5(1):9

Vaz E, Shaker RR, Cusimano MD (2020b) A geographical exploration of environmental and land use characteristics of suicide in the greater Toronto area. Psychiatry Res:112790

Verhoef ET, Nijkamp P (2004) Spatial externalities and the urban economy. In: Capello R, Nijkamp P (eds) Urban dynamics and growth, advances in urban economics. Elsevier, Amsterdam

Xiaoying L, Xiaowen L, Wanglu P, Tong C (2004) Modelling urban sprawl with the optimal integration of Markov chain and spatial neighborhood analysis approach. In: Geoscience and Remote Sensing Symposium, 2004. IGARSS'04. Proceedings. 2004 IEEE International, vol 4. IEEE, Anchorage, pp 2658–2661

Chapter 8
Spatial Association of Agricultural Land Loss in Southern Europe

Abstract This chapter looks at the integrative complexity of regional land use change in southern Europe. Examining the locations of agricultural and urban hotspots and land use interactions throughout southern Europe, a spatial accounting methodology is adopted to understand what changes and interactions led to the unprecedented impacts of land use change. In the first instance, the impacts of land use change in southern Europe are assessed by quantifying the loss of agricultural land, followed by the effects these changes had at the regional level. These issues are extended within a spatial decision support systems framework in line with the main drivers of future sustainable choices for regional sustainable development in southern Europe. The usage of local spatial autocorrelation techniques allows us to determine hotspots of the impact of significant clusters of spatial land use change. This calls out for the importance of local policies to consider the relevance of rural habitats and small towns, leading to an integrative perspective of the importance of geography as a planning instrument for assessing and monitoring southern Europe's sustainable future.

Keywords Land use change · Southern Europe · Regional impact dynamics · Local spatial autocorrelation

8.1 Southern Europe: A Region in Unprecedented Transition

The world has witnessed in recent years unprecedented transformation concerning land cover and land use change (Fuchs et al. 2015). This has accumulated ongoing concerns from a land and urban planning standpoint (Couch et al. 2007; Vaz et al. 2012). Policies towards resource management should foster clear sustainable guidelines (Berkes 1989). Land use transitions, however, seem to pose conflicting avenues concerning what would be a harmonious regional development because of depletion of certain land use types (Li et al. 2015). These conflicting visions are chiefly a result of different policies and needs of anthropogenic activity, culminating on the regional

E. Vaz, T. de Noronha, *Sustainable Development in Southern Europe*,
https://doi.org/10.1007/978-3-662-62177-6_8

discrepancies of diverse social, economic, and environmental demands and challenges of future growth. Accumulated by the gradual change due to urbanization, where acceleration of urban growth jeopardizes fragile environments, anthropogenic activity must consider the effects on derelict agriculture, natural environment, heritage, and the sustainability of ecosystems (Antrop 2004). Countries lacking continuous policy programs are particularly experiencing significant issues pertaining unmanned land use change (Force et al. 1999), given the interest of economic stakeholder's demand on the carrying capacity of already depleted regions impacting ecology, landscape, and biodiversity (Geoghegan et al. 1997). In developing countries, this is particularly witnessed within the rapid population increase in urban settlements leading to the requirement of a prompt response from stakeholders to cope with urban pressure (Ahluwalia et al. 1979; Arsanjani et al. 2013b; Kraas et al. 2013; Onilude and Vaz 2020; Vaz 2014). In developed countries, however, the recent economic recession in Europe has brought the need of Europe's panorama to rethink itself at the regional level given a series of concerns that may impact employment for future generations, particularly in southern Europe (Goldstein et al. 2013). The disparity in Europe in terms of economic performance seems to have underlying historical roots in the different regional performance (Tabellini 2010) suggesting that policy instruments as well as local analytics must be used as planning tools (Adams 2016). In the context of historical land use change, urbanization had such an extensive impact on agricultural land use change (Goldewijk 2001), extending to the regional impacts of agricultural land loss in the last decades (van Vliet et al. 2015). The changes in agricultural land witnessed in southern Europe have been manifested by a significant deintensification in peri-urban areas (Vaz et al. 2014), leading to unequaled challenges at land use level, further heightened by the Great Recession. Currently, policy making and decision-making in southern Europe must thus create mechanisms to address the challenges of regional asymmetries, particularly with the materialization of brownfields because of obsolete infrastructures (Loures and Vaz 2016). The need for southern Europe to reinvent its role in positioning itself as a key driving force within Europe's engine of economic growth strongly urges for an integration of its heritage and landscape (Bandarin and Van Oers 2012), as well as an integration of its rural spaces (Hoggart et al. 2014).

The growing asymmetries and posing socioeconomic challenges such as aging and migration (King 2000) bring significant issues to the development of agricultural regions. These asymmetries at Europe's regional level are a direct result of one of the greatest economic recessions of the modern history and have brought serious consequences on priorities of southern Europe's sustainability and of its economic climate (Matsaganis and Leventi 2014). Many urban areas have become inactive, and the burden of maintaining depreciated tourism infrastructures leads to the encroachment of agricultural, forest, and natural areas within cities. The loss of functional usage of these infrastructures in urban environments sets out the stage for potential brownfields (Nijkamp et al. 2002) that can however be measured at the regional level throughout the means of spatial analysis and geographic information systems (Thomas 2002). Furthermore, the loss of rural habitats leads to land desertification increased by the effects of expected climate change (Krimly et al.

2016), where several of southern European urban littoral regions may be irreversibly lost (Hallegatte et al. 2011). In this sense, it is of utmost importance to understand the effects of land use types of natural and urban habitats as well as the existing transactions of socioeconomic impacts of land use dynamics (Lambin and Meyfroidt 2010).

Tackling these changes can only be carried out by a systematic monitoring of quantitative analytics at the regional level so as to understand the stagnant nature of change in which the spatial component is of paramount importance (Fischer and Getis 2009). The analytical understanding at the regional level of the traction of land use change leads thus to better policies that can prevent further damages on agricultural land use at the regional level. Methodologies that offer integrated extractions of land use categories and quantify the change using different classification techniques venturing from remote sensing and land use change models are an intrinsic part of a more sustainable regional future (Arsanjani et al. 2013a). The availability of data through the different gradients for regional development indicators at the European level, as well as the technological advance of an increasing amount of volunteered geographic information, allows to abridge the national level concerns of countries in southern Europe with a strengthening of regional and local level analytics (Coelho et al. 2010). Despite the significant potential of land use analysis at the regional level, computational advance to understand larger regions is recent (Arsanjani et al. 2016). Thus, only a limited number of studies are at present available understanding the entire regional challenges faced throughout the world. However, with the current advances of geocomputation, land use dynamics combined with spatial analysis and geostatistics can share a significant understanding of agricultural transitions for larger extents (Bateman et al. 2013; Lawler et al. 2014; Prishchepov et al. 2017). In this sense, this chapter focuses on the regional spatiotemporal aspects of land use transitions in southern Europe. We attempt to (i) detect and quantify the dynamics of agricultural land use change, (ii) incorporate spatial autocorrelation as to understand the differences at the regional level of the traction of these changes, and (iii) understand how spatial decision support system of the present challenges of land use transitions in southern Europe's most fragile agricultural regions.

8.2 Data and Methods

8.2.1 Data

The following countries were included in the study: Portugal, Spain, Italy, and Greece. CORINE Land Cover (CLC) data regarding land use change was downloaded from http://land.copernicus.eu/. CLC was started in 1985 as an attempt for joint coordination of land use data and to address common goals of several environmental issues. The CLC datasets and several resulting programs are linked to the European Environment Agency, fomenting the global emphasis of the different

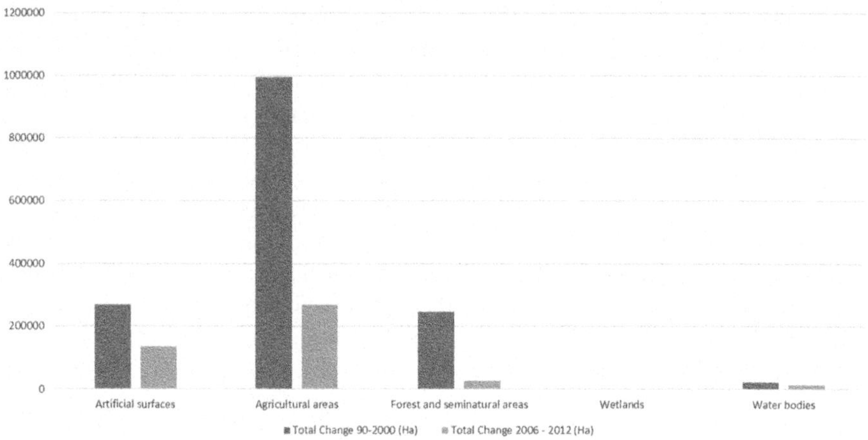

Fig. 8.1 Land use change registered between CLC 90 and 00 and CLC 00 and 06

programs through standardized approaches, particularly in regard to monitoring and inventorying land cover (Heymann 1994). In the CLC program, now in its fourth edition with community available sets for 1990, 2000, and 2012, the databases have systematically contained 44 land cover classes and presented as a cartographic product easy to download through different national repositories at a scale of 1:100 000.

Within the umbrella of the European Commission, the program has become a leading repository for Europe's land use data. The used CORINE Land Cover (CLC) data allows for a regional change assessment (Falcucci et al. 2007; Feranec et al. 2017) over the multiple timestamps available from the original datasets CLC90 (1985), CLC00 (2000), CLC06 (2006), and CLC12 (2012) (Soukup et al. 2016; Mellino and Ulgiati 2015). The datasets were provided as vector layers and were consequentially clipped into the southern Europe area. This consisted of the initial step of data preparation, allowing to integrate the different data components. Further, agricultural land use data was measured within the transaction of land use change, by means of compiling the different datasets between the changing time series. This allowed for a multi-temporal comparison of land use change (Fig. 8.1).

8.2.2 *Defining Agricultural Hotspots*

A hotspot corresponds traditionally to a concentration of incidents of any geographical structure within a given spatially explicit boundary (Getis and Ord 1992). The importance of understanding the existence of hotspots from a land use perspective is significant. Hotspots integrate the burden of carrying capacity over land, and, on the other, influence of hotspots over time may lead to a clearer understanding of the tendency of a given spatiotemporal pattern over time (Verburg et al. 2006). Hotpots

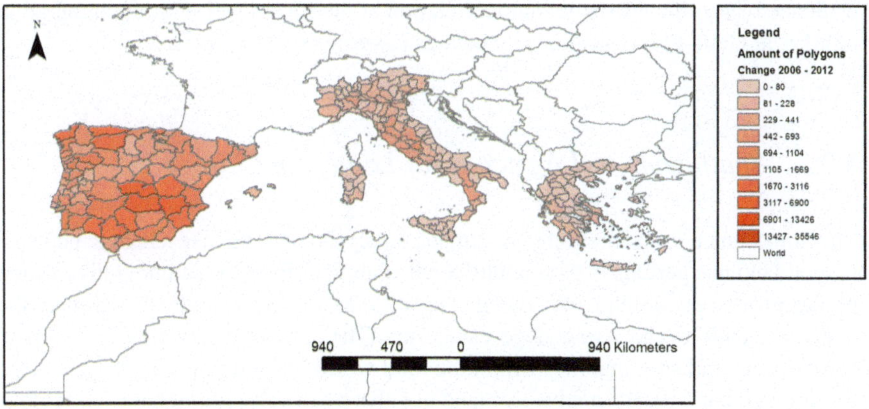

Fig. 8.2 Quantification of land use change centroids at NUTS III level between 2006 and 2012

express due to their geovisual and spatiotemporal characteristics the dynamics of change that from a policy perspective can support land use planning (Rounsevell et al. 2006). The concentration thus of certain patterns within geographical structures permits a better management of agricultural land (Wang et al. 2016), where hotspots of urban increase in comparison to loss of agricultural land are important to measure within the sustainability of ecoregions, sustainable landscapes, and integrated urbanization processes (Tan et al. 2005). In the case of the significant change of hectares between the last two decades in CORINE Land Cover, a better understanding of the location of these changes is of utmost importance for planning structures in southern Europe, where incentives for agricultural and rural activities, given the intrinsic abandonment of rural land, must be carefully equated and designed within the challenges of regional dynamics in the coming decades. Combat to desertification and sustainable development must be carried out by an adequate integration of macro- and microscale analysis of policy reforms mitigating territorial disparities (Renwick et al. 2013), particularly in southern Europe where land degradation is significantly increasing (Salvati and Zitti 2007). Bringing spatial analytical methods allows to infer a gradient of concentration at the regional level, particularly pertaining to the dynamics of land use while constructing new strategies for abandoned regions which have become unproductive for agricultural use and thus are prone to land degradation. Figure 8.2 shows the loss of agricultural land for built-up land use types, through quantification of the change centroids registered between 2006 and 2012 and NUTS III level.

The distribution of land use change within 2006 and 2012 clearly shows the existence of underlying increase of land use change within the Iberian Peninsula, as well as the north-western coast of Italy. Understanding the spatially explicit change dynamics is also recurrent in coastal areas, justifying a closer analysis as well as quantification of these regions. This regional assessment is done by performing a Global Moran's I indicator for the region, and further hotspots are assessed after evidence of significant spatial autocorrelation at the local level with the integration of a Local Getis-Ord (Local Gi*) indicator (Overmars et al. 2003).

8.2.3 *Spatiotemporal Agricultural Land Use Change*

The importance of addressing systematic analysis of land use change between several land use periods is paramount. Such analytics allow for an integrative vision of the processes of land use change, as well as a clearer understanding of the dynamics within land use change categories. The important policies of fostering the diversity and sustainability within the rural world make it of particular relevance to address the composite analysis of land use change of agricultural land. To achieve this, net change was reported between the time frames of 1990–2000 and 2006–2012, based on the retrieved CORINE Land Cover data inventory. The advantage of using CORINE Land Cover pertains to the existence of the same cell size and resulting spatial resolution. This allows for the construction of a cell row i–column j which presents the number of cells experiencing a transition from type i to type j in the corresponding area. This allows to quantify the spatial transformation for each parcel, leading to a better understanding of the spatial configuration within the landscape, consequentially assessed in a GIS environment. The types of land use changes are assembled into the calculated Local Getis-Ord score and permit the quantification of the succession of land use types, for example, $i \rightarrow j$ or $i \rightarrow j \rightarrow i$.

8.3 Discussion

8.3.1 *Agricultural Land Use Change*

Adequate interpretation of land use change is of great importance for successful regional planning. With the advent of a significant amount of change detection techniques for regional science, novel computation methods have become available to better assess these changes. In the case of agricultural land use types, the literature is still quite scarce. One of the main advantages of integrated spatial analysis for agriculture is the potential for long-term planning of ecological futures. This allows for protection of integrated sustainable landscapes, combined with the social responsibility as well as the commitment to transmit our heritage to future generations by keeping sustainable development. The agricultural environment is of utmost importance in this endeavor, particularly the rural habitats which allow small localized farmers to subsist. A quantitative assessment of the five most significant agricultural land use class changes into continuous and discontinuous urban fabric is shown below (Figs. 8.3 and 8.4). A significant amount of nonirrigated arable land (43%) was lost to continuous urban fabric, followed by 23% of loss of complex cultivation

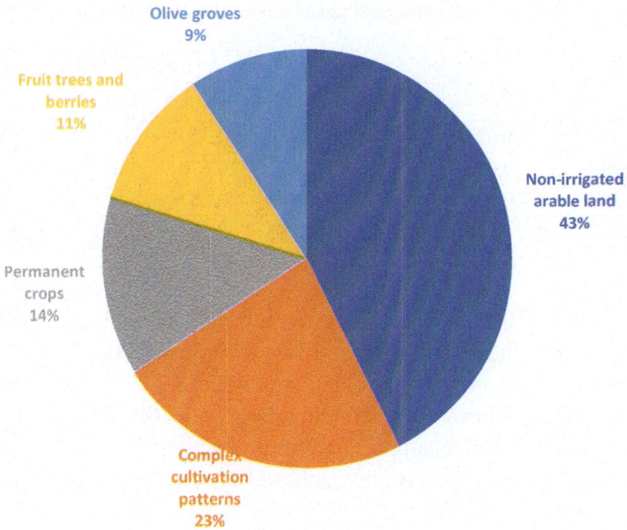

Fig. 8.3 Agricultural land loss to continuous urban fabric

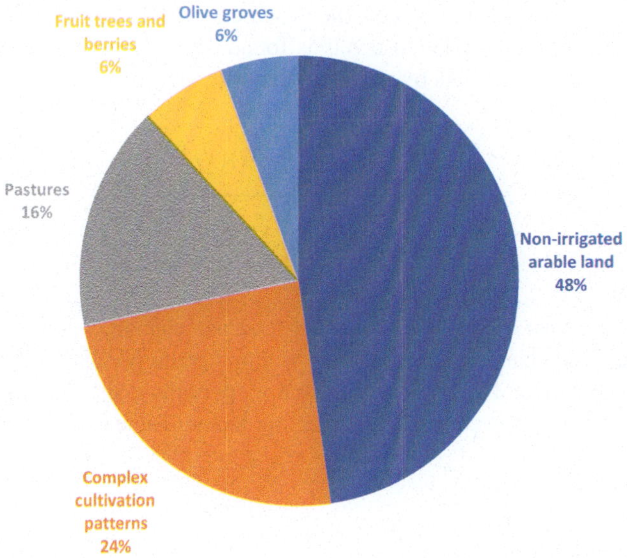

Fig. 8.4 Agricultural land loss to discontinuous continuous urban fabric

patterns, corresponding to small farmers with mixed agricultural land use types (Fig. 8.3). Furthermore, it is important to note that olive groves are an important heritage of southern Europe, of which between 2006 and 2012, 8% were lost (Fig. 8.3). For discontinuous urban fabric, the patterns are similar in southern Europe (Fig. 8.4). Nonirrigated arable land loss has increased by 5%, summing a total of

Table 8.1 Moran's index for global spatial association in Southern Europe

Country	Moran's index	Expected index	Variance	z-score	p-value	Distance threshold (km)	Projection (WGS84 UTM zones)
Portugal	0.153	−0.025	0.093	0.584	0.559	66.18	28N
Spain	0.073	−0.012	0.039	0.426	0.670	148.02	29N
Italy	0.071	−0.006	0.030	0.444	0.657	140.56	31N
Greece	0.099	0.099	0.036	0.583	0.560	120.26	34N

48% of loss, while classes such as olive groves, fruit trees and berries, and permanent crops registered a slightly less significant loss for urban fabric. Regarding complex cultivation patterns, an increment of 1% was noticed for land use change within discontinuous urban fabric, corresponding largely to the additions of rural and agricultural infrastructures to support agricultural activity.

No significant spatial association was found at the global level for any of these countries, inferring that the distribution of land use change of agricultural land use does not target any specific regions, but targets randomly. The lack of a strong spatial association at the global level is of particular concern, as it suggests that there has not been at the national level the existence of planning tools to measure the impacts of land use change, and lack of a systematic vision that could allow a better assessment of nonrandom spatial association levels is found. The Global Moran's was tested with inverse distance weights by means of the following calculation:

$$I = \frac{N\sum_{i=1}^{n}\sum_{i=1}^{n} w_{ij}(x_i - \overline{x})\left(x_j - \overline{x}\right)}{\left(\sum_{i=1}^{n}\sum_{j=1}^{n} w_{ij}\right)\sum_{i=1}^{n}(x_i - \overline{x})^2} \tag{8.1}$$

where N is the total count of centroids of land use change, $\overline{x}$ is the mean of the observed centroid changes, x_i is the value within a given NUTS III region, x_j is the value of another given NUTS III region, and w_{ij} pertains to the Euclidean distances between centroids that were used as weighting factor (Table 8.1).

8.3.2 Local Spatial Autocorrelation in Southern Europe

The fine-grained information available by means of spatially explicit repositories allows for integration of socioeconomic data within the decision-making structure. Geographic information systems have allowed to distill emergent behaviors through incorporation of locational information. This allows for a high spatial resolution assessment of underlying dynamics of anthropogenic behavior on land use, landscape, and ecosystems in what are defined as complex spatial systems. Nested in the

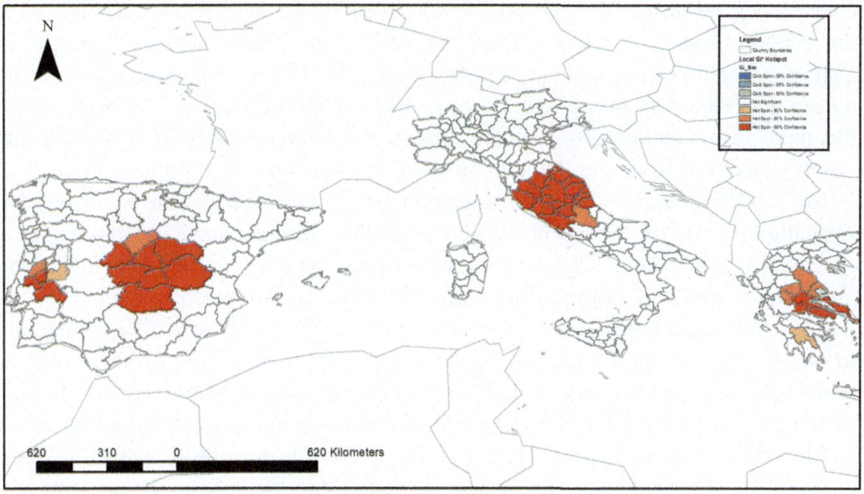

Fig. 8.5 Location of Local Gi* hotspots in southern Europe

importance of regional decision processes, the bottom-up approach of local data leads to the potential of creating better planning structures that may enhance from the local planning sphere regional and global structure. These deterministic models may use quantitative methods to robust and strengthen local policy and support the nonlinear transitions of changing land use at the global level. This interaction is fostered by GIS that offers a set of tools that interpret the spatial components over complex dynamics of regions (Fischer and Getis 2009). The spatial resolution of NUTS III is provided useful for this analysis. The ongoing construct of aggregated data that holds enough detail to visualize local-level land use is provided useful to account for spatial differences between NUTS III levels. By means of spatial autocorrelation, we can understand the nature and characteristics of location dynamics (Getis 2010). A Getis-Ord algorithm was performed indicating significant hotspots of concentration of agricultural land loss in southern Europe. The resulting hotspot analysis, carried out with ArcGIS 10.3, brought a Gi* statistic in the form of a z-score, indicating the number of regions with either high or low value of spatial clusters. Statistically significant clusters coincide with a z-score greater than 1.96 (Mitchell 2005) as observed in Fig. 8.5.

8.4 Discussion

Spatial analysis has become an important instrument for regional planning, subdued to the availability of data and interpretation of results fostered by policy makers and stakeholders even when considering imprecise socioeconomic and spatial determinants (Leung 2013). Rich data analytics and the integration of spatially explicit datasets help to design the choices of more sustainable and harmonious choices at the geographical level (Nagendra et al. 2004). Oftentimes, however, local-level analytics seem to be neglected, and only global indicators are not enough to generate a deep understanding of regional problems. In southern Europe, it has been shown that there are evident hotspots that are culprits of the loss of agricultural land use that deserve local attention and the integration of regional and local policies (Vaz et al. 2015). It is these regions, however, nested in the central regions of the countries, that agricultural land has predominantly changed into different land use typologies – specifically urban land. The lack of tailor-made policies accrues the increasing pressure on southern Europe rural areas, indicating a strong connection with the rise of urban metropolitan regions in southern Europe, particularly in areas that aim to become significant urban hubs in the service sector. This has a direct effect of urban sprawl on the ongoing sustainability of the rural world in these growing urban hubs that must envision to become sustainable cities and consider the peri-urban environments (Camagni et al. 1998). While the transition from an industrial-based model in most of southern Europe has gravitated into the service sector (Linz and Stepan 1996), with the ongoing economic recession in these countries, the loss of agricultural and rural areas may have unprecedented consequences on the suitability of rural land in the future. It is of utmost importance that the rural world considers the transition of natural, social, and economic development as key drivers for sustainable development, but aggregates these toponymical dimensions within the roles of space, location, and place (Fig. 8.6).

The relation of these dimensions shares crucial data that may allow local and regional models to indicate the traction of data into information, procuring through spatial analysis and modelling the generation of applied knowledge structures that may change the dynamics of a bottom-up approach of regional development. In the case of southern Europe, where central regions are losing rapidly agricultural land use types to urban concentration, this is of paramount importance, as loss of sustainable agricultural land and rural heritage may well be irreversible.

8.5 Conclusions

As much as our impact on Earth has brought irreversible environmental change, our landscapes have in detriment of these choices witnessed a substantial change, most of it affecting our natural and rural areas. In the context of regional agricultural development, economic geography and complex space-time dynamics are factors of continuous change in southern Europe. Monitoring of the transitions of land at the regional level is thus of utmost importance for sounder regions in the future (Jain et al. 2016). It is relevant to preserve landscapes by enabling efficient economic growth, without jeopardizing the natural ecosystems and mitigating the impacts on rural land and landscapes alike. This calls out for a new approach taking advantage of technological efficiency and information with local degrowth (Schneider et al. 2010). This chapter has shown the possible spatial interpretations of agricultural land

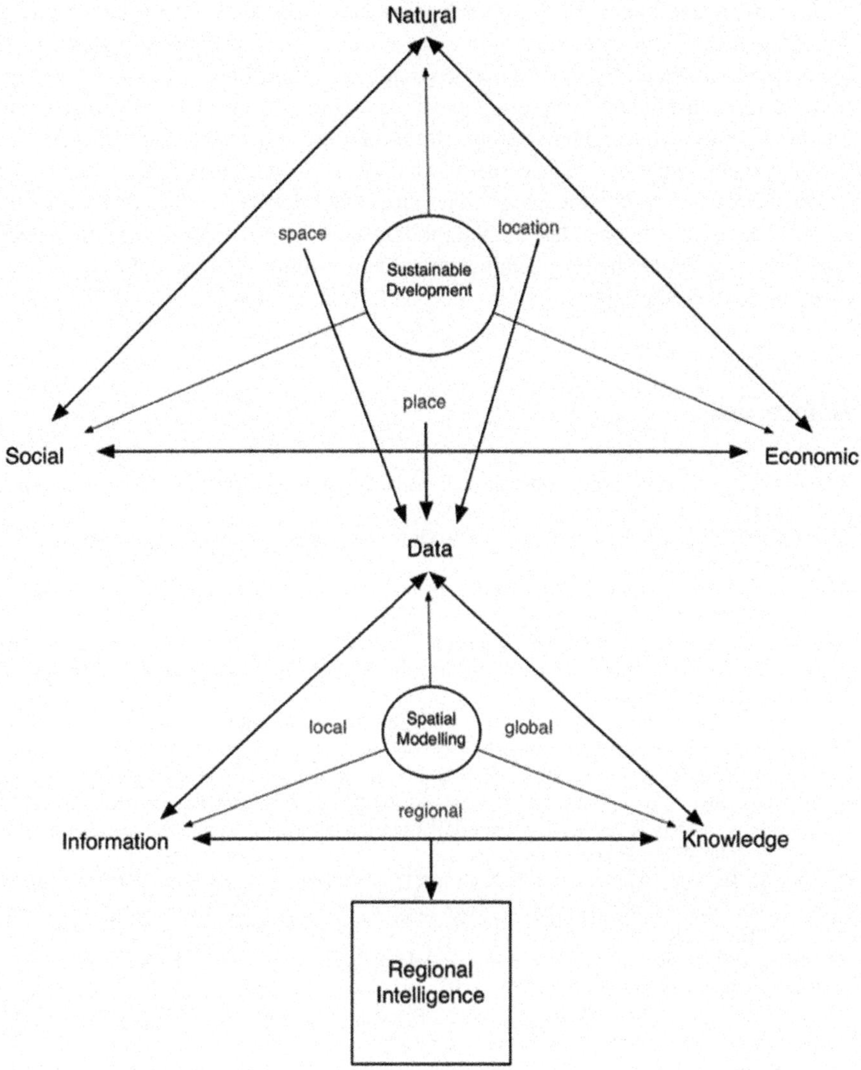

Fig. 8.6 The spatial dimension of sustainable development

use change by means of defining the role at present of GIS as tools to allow sounder urban and regional interactions. The proposed three pathways integrating regional development within a spatial landscape preservation framework should serve as a guideline for policy makers and regional planners to note future development of regions under rapid change, whether economic, social, or environmental (Willemen et al. 2010). Functional aspects of the landscape are therefore an utmost important trade of the future directions of our regions, relying on the memory of economic, social, and environmental transitions (Schulze and Gerstberger 1994).

The incremental concern of the collapsing landscape calls out for local policies and assessment to mitigate the continuous impact on rural habitats and small towns (de Noronha and Vaz 2015). From a geographical analysis perspective, regions can only become sustainable when agricultural and rural spatial memories, that is, the identity of place and time and economic traditions, are coherent and long lasting. The regional complexity of urban regions calls for a holistic perspective where rural landscapes must consider the integration of local decision support systems within the regional development agendas. Spatial models have the power to assess and support decisions to integrate better planning instruments within the regional development tools for southern Europe.

References

Adams N (2016) Regional development and spatial planning in an enlarged European Union. Routledge, London

Ahluwalia MS, Carter NG, Chenery HB (1979) Growth and poverty in developing countries. J Dev Econ 6(3):299–341

Antrop M (2004) Landscape change and the urbanization process in Europe. Landsc Urban Plann 67(1):9–26

Arsanjani JJ, Helbich M, Kainz W, Boloorani AD (2013a) Integration of logistic regression, Markov chain and cellular automata models to simulate urban expansion. Int J Appl Earth Observ Geoinf 21:265–275

Arsanjani JJ, Helbich M, de Noronha Vaz E (2013b) Spatiotemporal simulation of urban growth patterns using agentbased modeling: the case of Tehran. Cities 32:33–42

Arsanjani JJ, Tayyebi A, Vaz E (2016) GlobeLand30 as an alternative fine-scale global land cover map: challenges, possibilities, and implications for developing countries. Habitat Int 55:25–31

Bandarin F, Van Oers R (2012) The Historic Urban Landscape: managing heritage in an urban century. John Wiley & Sons, Chichester

Bateman IJ, Harwood AR, Mace GM, Watson RT, Abson DJ, Andrews B et al (2013) Bringing ecosystem services into economic decision-making: land use in the United Kingdom. Science 341(6141):45–50

Berkes F (1989) Common property resources: ecology and community-based sustainable development. Belhaven Press, London

Camagni R, Capello R, Nijkamp P (1998) Towards sustainable city policy: an economy-environment technology nexus. Ecol Econ 24(1):103–118

Coelho P, Mascarenhas A, Vaz P, Dores A, Ramos TB (2010) A framework for regional sustainability assessment: developing indicators for a Portuguese region. Sustain Dev 18(4):211–219

Couch C, Leontidou L, Petschel-Held G (2007) Urban sprawl in Europe. Blackwell Publishing Ltd, Oxford

de Noronha T, Vaz E (2015) Framing urban habitats: the small and medium towns in the peripheries. Habitat Int 45:147–155

Falcucci A, Maiorano L, Boitani L (2007) Changes in land-use/land-cover patterns in Italy and their implications for biodiversity conservation. Landsc Ecol 22(4):617–631

Feranec J, Soukup T, Taff GN, Stych P, Bicik I (2017) Overview of changes in land use and land cover in Eastern Europe. In: Land-cover and land-use changes in Eastern Europe after the collapse of the Soviet Union in 1991. Springer International Publishing, Dordrecht, pp 13–33

Fischer MM, Getis A (eds) (2009) Handbook of applied spatial analysis: software tools, methods and applications. Springer Science & Business Media, Berlin

Force UT, Britain G, Rogers RG (1999) Towards an urban renaissance. Spon, London

Fuchs R, Herold M, Verburg PH, Clevers JG, Eberle J (2015) Gross changes in reconstructions of historic land cover/use for Europe between 1900 and 2010. Glob Change Biol 21(1):299–313

Geoghegan J, Wainger LA, Bockstael NE (1997) Spatial landscape indices in a hedonic framework: an ecological economics analysis using GIS. Ecol Econ 23(3):251–264

Getis A (2010) Spatial autocorrelation. In: Fischer MM, Getis A (eds) Handbook of applied spatial analysis: software tools, methods and applications. Springer, Berlin, pp 255–278

Getis A, Ord JK (1992) The analysis of spatial association by use of distance statistics. Geogr Anal 24(3):189–206

Goldewijk KK (2001) Estimating global land use change over the past 300 years: the HYDE database. Glob Biogeochem Cycl 15(2):417–433

Goldstein J, Kreyenfeld M, Jasilioniene A, Örsal DDK (2013) Fertility reactions to the "Great Recession" in Europe: recent evidence from order-specific data. Demogr Res 29:85–104

Hallegatte S, Ranger N, Mestre O, Dumas P, Corfee-Morlot J, Herweijer C, Wood RM (2011) Assessing climate change impacts, sea level rise and storm surge risk in port cities: a case study on Copenhagen. Clim Change 104(1):113–137

Heymann Y (1994) CORINE land cover: Technical guide. Office for Official Publication of the Europe Communities, Luxembourg

Hoggart K, Black R, Buller H (2014) Rural Europe. Routledge, Abingdon

Jain M, Dawa D, Mehta R, Dimri AP, Pandit MK (2016) Monitoring land use change and its drivers in Delhi, India using multi-temporal satellite data. Model Earth Syst Environ 2(1):1–14

King R (2000) Southern Europe in the changing global map of migration. In: Eldorado or Fortress? migration in Southern Europe. Palgrave Macmillan, London, pp 3–26

Kraas F, Aggarwal S, Coy M, Mertins G (eds) (2013) Megacities: our global urban future. Springer Science & Business Media, Dordrecht

Krimly T, Apfelbeck J, Huigen M, Dabbert S, Reichenau TG, Lenz-Wiedemann VI et al (2016) Effects of future climate changes on yields, land use and agricultural incomes. In: Regional assessment of global change impacts. Springer International Publishing, Dordrecht, pp 609–614

Lambin EF, Meyfroidt P (2010) Land use transitions: socio-ecological feedback versus socio-economic change. Land Use Policy 27(2):108–118

Lawler JJ, Lewis DJ, Nelson E, Plantinga AJ, Polasky S, Withey JC et al (2014) Projected land-use change impacts on ecosystem services in the United States. Proc Natl Acad Sci 111(20): 7492–7497

Leung Y (2013) Spatial analysis and planning under imprecision. Elsevier, Amsterdam

Li Y, Li Y, Westlund H, Liu Y (2015) Urban–rural transformation in relation to cultivated land conversion in China: implications for optimizing land use and balanced regional development. Land Use Policy 47:218–224

Linz JJ, Stepan A (1996) Problems of democratic transition and consolidation: Southern Europe, South America, and post-communist Europe. JHU Press, Baltimore

Loures L, Vaz E (2016) Exploring expert perception towards brownfield redevelopment benefits according to their typology. Habitat Int 45:72–81

Matsaganis M, Leventi C (2014) The distributional impact of austerity and the recession in Southern Europe. South Eur Soc Polit 19(3):393–412

Mellino S, Ulgiati S (2015) Monitoring regional land use and land cover changes in support of an environmentally sound resource management. In: Sustainable development, knowledge society and smart future manufacturing technologies. Springer International Publishing, Dordrecht, pp 309–321

Mitchell A (2005) The ESRI Guide to GIS analysis, Volume 2: spatial measurements and statistics. ESRI Press, Redlands

Nagendra H, Munroe DK, Southworth J (2004) From pattern to process: landscape fragmentation and the analysis of land use/land cover change. Agricul, Ecosyst Environ 101(2):111–115

Nijkamp P, Rodenburg CA, Wagtendonk AJ (2002) Success factors for sustainable urban brownfield development: a comparative case study approach to polluted sites. Ecol Econ 40(2): 235–252

Onilude O, Vaz E (2020) Data analysis of land use change and urban and rural impacts in Lagos state, Nigeria. Data 5(3):72
Overmars KP, De Koning GHJ, Veldkamp A (2003) Spatial autocorrelation in multi-scale land use models. Ecol Model 164(2):257–270
Prishchepov AV, Müller D, Baumann M, Kuemmerle T, Alcantara C, Radeloff VC (2017) Underlying drivers and spatial determinants of post-Soviet agricultural land abandonment in temperate Eastern Europe. In: Land-cover and land-use changes in Eastern Europe after the collapse of the Soviet Union in 1991. Springer International Publishing, Dordrecht, pp 91–117
Renwick A, Jansson T, Verburg PH, Revoredo-Giha C, Britz W, Gocht A, McCracken D (2013) Policy reform and agricultural land abandonment in the EU. Land Use Policy 30(1):446–457
Rounsevell MDA, Reginster I, Araújo MB, Carter TR, Dendoncker N, Ewert F et al (2006) A coherent set of future land use change scenarios for Europe. Agricult, Ecosyst Environ 114(1):57–68
Salvati L, Zitti M (2007) Territorial disparities, natural resource distribution, and land degradation: a case study in southern Europe. Geojournal 70(2-3):185–194
Schneider F, Kallis G, Martinez-Alier J (2010) Crisis or opportunity? economic degrowth for social equity and ecological sustainability. Introduction to this special issue. J Clean Prod 18(6): 511–518
Schulze ED, Gerstberger P (1994) Functional aspects of landscape diversity: a Bavarian example. In: Biodiversity and ecosystem function. Berlin Heidelberg, Springer, pp 453–466
Soukup T, Feranec J, Hazeu G, Jaffrain G, Jindrova M, Kopecky M, Orlitova E (2016) CORINE land cover 1990–2000 changes: analysis and assessment. European Landscape Dynamics: CORINE Land Cover Data, p 13
Tabellini G (2010) Culture and institutions: economic development in the regions of Europe. J Eur Econ Assoc 8(4):677–716
Tan M, Li X, Xie H, Lu C (2005) Urban land expansion and arable land loss in China—a case study of Beijing–Tianjin–Hebei region. Land Use Policy 22(3):187–196
Thomas MR (2002) A GIS-based decision support system for brownfield redevelopment. Landsc Urban Plann 58(1):7–23
van Vliet J, de Groot HL, Rietveld P, Verburg PH (2015) Manifestations and underlying drivers of agricultural land use change in Europe. Landsc Urban Plann 133:24–36
Vaz E (2014) Managing urban coastal areas through landscape metrics: an assessment of Mumbai's mangrove system. Ocean & coastal management 98:27–37
Vaz E, De Noronha T, Nijkamp P (2014) Exploratory landscape metrics for agricultural sustainability. Agroecol Sustain Food Syst 38(1):92–108
Vaz E, Painho M, Nijkamp P (2015) Linking agricultural policies with decision-making: a spatial approach. Eur Plann Stud 23(4):733–745
Vaz E, Nijkamp P, Painho M, Caetano M (2012) A multi-scenario forecast of urban change: A study on urban growth in the Algarve. Landsc Urban Plann 104(2):201–211
Verburg PH, Schulp CJE, Witte N, Veldkamp A (2006) Downscaling of land use change scenarios to assess the dynamics of European landscapes. Agricult, Ecosyst Environ 114(1):39–56
Wang H, Qiu F, Ruan X (2016) Loss or gain: a spatial regression analysis of switching land conversions between agriculture and natural land. Agricult, Ecosyst Environ 221:222–234
Willemen L, Hein L, Verburg PH (2010) Evaluating the impact of regional development policies on future landscape services. Ecol Econ 69(11):2244–2254

Chapter 9
Southern Europe: The New Regional World

Abstract This chapter launches the comparative analysis of current land use patterns concerning the re-designation of agricultural land for urban use, permitting us to relate the results of urban variation per municipality with a variation of losses in RAN. The Algarve, characterized by a strong quality touristic activity, is a case study that suffers from a severe loss of agricultural areas and a significant reduction of the RAN. The recent European recession enhanced most of the agricultural areas to increasing abandonment to the detriment of the leapfrogging of peri-urban infrastructure. The double assessment of urban land and changes in the RAN has allowed understanding the impacts of both urban concentration in littoral areas and at spatial level, perceiving the changing aspects of land use in the Algarve. The municipal information for the Algarve also incorporates the option of understanding local-level impacts of such changes, such as the case of Silves, where the citric production is traditionally high in parallel to a current urban growth that follows that major trend of construction in peri-urban areas. Data sets of the RAN were summed to the same time frames as CLC data, and information was cross-linked. The results presented Alcoutim, Monchique, and Silves, with urbanist growth and explicit loss of agricultural land, by overstepping the RAN's agricultural regulation in the case of Silves. Of course, at the local level, this information is not evident. But a regional quantification of variations of the RAN and urban areas may offer much clarification of land-change patterns for the Algarve.

Keywords Agricultural land loss · RAN · Land-change patterns: Algarve · Entrepreneurial behaviors · Lagging areas · Southern Europe · Regional sustainability

9.1 Introduction

Considering the Network of Heads of European Environment Protection Agencies, 2005, one of the essential aspects of competitiveness is environmental regulation. It does not only make cost for industry and business more affordable but also creates new markets for environmental goods and services, some of them even with new

E. Vaz, T. de Noronha, *Sustainable Development in Southern Europe*,
https://doi.org/10.1007/978-3-662-62177-6_9

innovative technologies, thus generating new jobs and protecting scarce natural resources. Environmental law comprehends the environmental law system and should be defined as "an organized way of using all of the laws in our legal system to minimize, prevent, punish, or remedy the consequences of actions which damage or threaten the environment, public health, and safety". Sustainable development should be assessed from a multi-disciplinary lens, considering the distribution of resources. This is frequently overlooked, and as regulation is often overstepped due to economic factors, effective regulation becomes an area of disagreements between the paradigm of growth and the wish for sustainability. As a result, in recent decades, environmental deterioration and increasing economic growth have brought insufficiencies to certain ecological sectors, leading to increasing asymmetries, for example, in agriculture.

To "minimize the consequences on environment" becomes a very tough task, and it requires new regional rules to articulate those policies of sustainable development and environmental change, aside from the paradigm of socioeconomic growth. Legislation is, however, often restructured and reorganized to fit the current, multiple features of environmental degradation, and, thus, they lack a stable and continuous monitoring of sustainability. Environmental deprivation caused by human actions has been identified in many areas of the world. In Europe, for example, where, in overall, strong legislation and a good legal system prevails, urban sprawl has been unavoidable. Together with population increase and some socioeconomic growth in urban areas, a significant land abandonment is taking place, especially in regions where service activities are in higher demand. In southern Europe, such kind of pressures originating from intensive tourism activities, associated with the lack of awareness related to the long-term costs of devastation, has directly been responsible for fragile ecosystems, loss of agricultural land, and, all together, the increase in coastal vulnerability. Coastal regions in southern Europe share two distinct issues in regards to policy implementation: (i) as a socioeconomic system they are highly productive regions that rely on complex modelling to understand, and (ii) on the other hand, the productive cycle of such areas relies heavily on ecosystems' functionality which may be jeopardized by excessive exploitation of goods and services, as is the case of tourism, one of the most predominant activities along the Mediterranean cost.

From a historical viewpoint, coastal areas have been the support for a diversity of resources such as agriculture, leading to settlement patterns which have stimulated regional prosperity. Their unique sceneries often combined to pleasant temperatures, and a historic-cultural appeal led to the vast expansion of tourist industries in such areas. Thus, a careful planning of the ecosystem support services of littoral regions should be implemented as soon as possible in an integrated set of efforts not to damage the fragile ecological habitats in such areas. In fact, the topic is about the resilience of the environmental carrying capacity to sustain the permanent search for economic growth. Though, for example, tourism is a beneficial activity to some extent, as it creates jobs in coastal areas, the counterpart is rapid land deterioration as a result of seasonal population pressure. The synergetic relation between economic growth and sustainable development is a very intricate one. As a system, the effects of socioeconomic growth impact it in a nonlinear, sometimes unexpected, way. The dynamics of nonlinear complex systems are very hard to deal with and oblige to an intelligent and effective multidimensional legislation. Yet, such environmental legislation has less effect upon decision-making, while narrow and focused environmental laws have a greater impact, as their application is more explicit.

For a more effective analyses of environmental endangerment of coastal areas (and, of course, other specific geographic spaces), spatial analysis serves as the major tool. Firstly, it has been mostly driven by different scientific disciplines, such as geography, statistics, economics, and mathematics. Secondly, by integrating analytical methods of complex systems (a consensus regarding their definition is still lacking), it was possible to add structural analysis by the combination of factors of economic, social, and natural drivers. One of the main advantages of complex systems analysis resides in the possibility of having an integrated approach to better understand the global consequences of interactions. The accessibility to spatial information and higher spatial resolution georeferenced economic, social, and environmental strata permits a much more coherent approach to integrated analysis: social, economic, and environmental phenomena happening in an exact space and time. By merging different factors from heterogeneous variables that happen within a territorial unit over time, it becomes possible to discover a coherent explanation of the key drivers for environmental change through spatial metrics, facilitating the calculation of sustainable development. Thus, spatial information and complex systems may, if combined appropriately, produce suitable approaches to land use change, offering support in identifying the key drivers for many land use changes. The cross-linkage of policy decisions implies a direct impact on land use and on territorial management. This information permits a much more accurate approach to decision-making and for understanding the relevant constraints that affect sustainable development.

9.2 A Study Case: The Algarve

If possible, a complete study of coastal southern Europe would well illustrate the arguments of the previous point. However, such would take us most of the energy and time dedicated to this book. So, our decision was to observe attentively the earlier described factors by means of a unique and reliable example. We consider the choice of Algarve, located at the southernmost region of Portugal, as the specific case study and the best example we could have to follow a dynamic approach of land use change in southern Europe. Certainly, cases differ from each other, but many common traces can be found across this Mediterranean line, for example, their unique ecological landscape, mostly forming part of the continental network of conservation habitats, many of them defined under the European Union Directives: 79/409/CEE and 92/43/CEE.

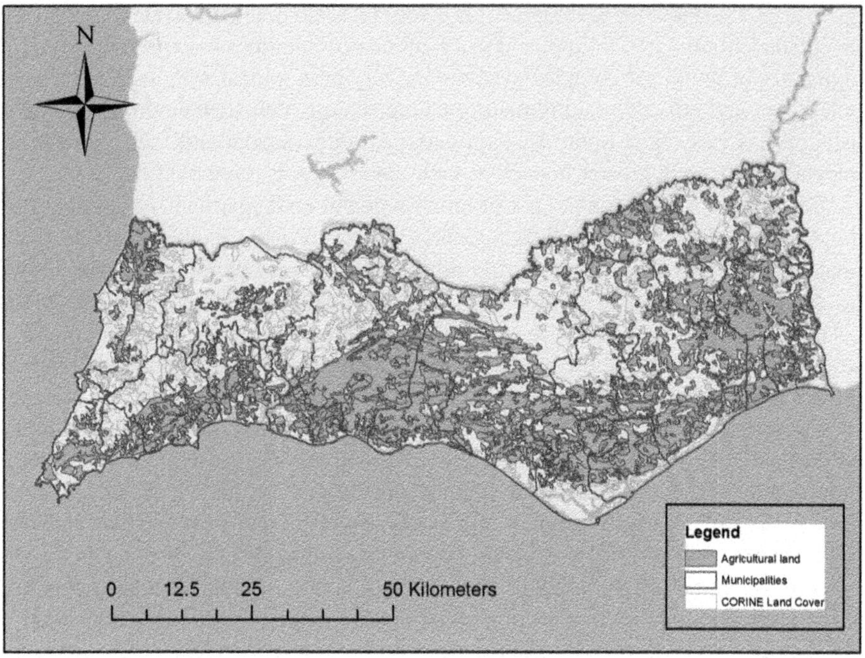

Fig. 9.1 Location of protected areas within the municipalities of the Algarve

The geomorphology of the Algarve is generally classified into three different groups: the *Interior*, the *Barrocal*, and the *Litoral*. Although they all differ, there is a substantial asymmetry between the *Interior* (located at the north of the region) and the *Litoral* (the coastal line of the Algarve).

Figure 9.1 represents the geographical region of the Algarve and inside it those parts of land which belong to NATURA 2000 network. The increasing imbalances between population increases in the south of the Algarve when compared with its decay in the north are threatening the important ecosystems and risking the agenda for development of rural areas.

Whereas agricultural activity has been substantial in the *Barrocal*, basically carob production, and in the *Interior*, where the activities are of agropastoral nature, the *Litoral* has mostly lost its farmed activities to the exploration of tourism and its related actions. Since the 1960s, the rise of tourism enhanced by intensification of low-cost flights throughout Europe has been a fast opportunity for economic growth and prosperity of the region. The creation of services and infrastructures to back up a massive tourism industry changed the nature of the activities from the primary sector into the tertiary sector, focusing preferentially on all subsectors related to tourism. And yes, the development of the tourist industry has provided better job opportunities, attracted massive population to the region, and contributed directly to coastal population increase. Still today, when tourism is booming again, the Algarve lacks

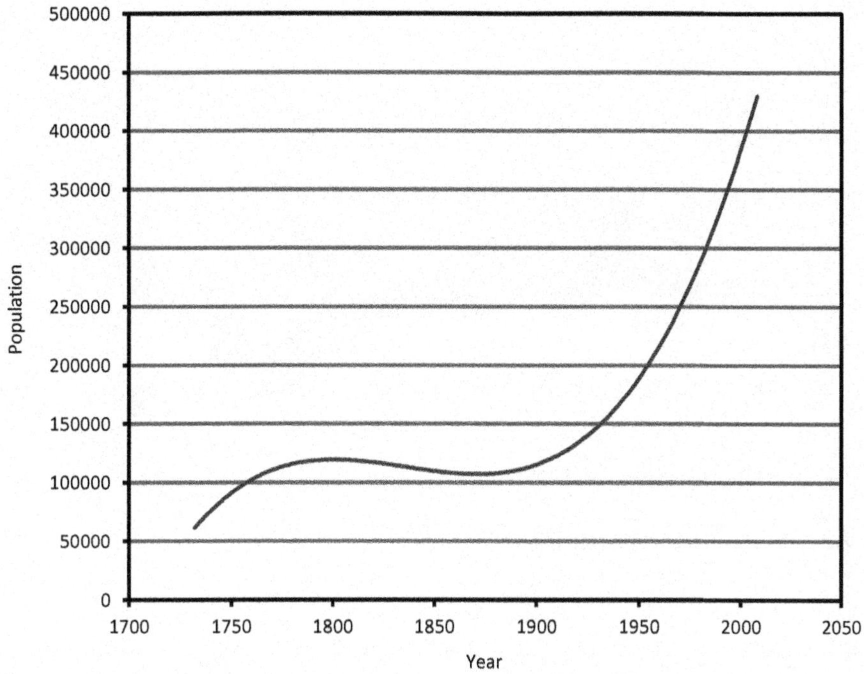

Fig. 9.2 Population growth in the Algarve since the seventeenth century

skills to face the need of hotel industry and restauration. Figure 9.2 confirms this trend in terms of population growth since the mid-seventeenth century.

The Algarve experienced an exponential growth, particularly since the 1980s, as a result of mass tourism industry. The first increase in the population took place during the eighteenth century and was a direct result of the productivity of the local fishing industry, which provided jobs and economic prosperity during the eighteenth and nineteenth centuries. The next booming phase took place from the 1970s (in 1973, there was 63,682 inhabitants) to 1990s. In 1992, the population growth rate was 167.62 percent, bringing the total population to 411,468, in 2004. This significant rise is directly related to a new type of economic growth resulting from the development of low-cost flights.

In 2008, the Algarve had a density of around 80 inhabitants per km^2. The asymmetry between the coastal area and the northern area of the Algarve escalates and exacerbates during the summertime when its population triplicates, clustering in the areas of tertiary sectors. Figure 9.3 shows the population density per parish, clearly reflecting the pattern of clusters along the coastal areas and lower densities in the interior. The areas with the highest population density are detected in the surroundings of Faro, the district capital of the Algarve, and in the areas of Albufeira, Portimão, and Vila Real de Santo António.

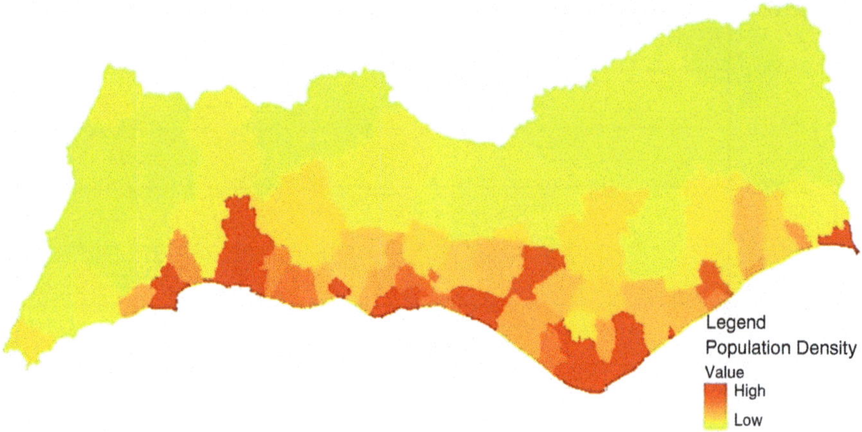

Fig. 9.3 Population density per parish

9.3 Data and Methodology

The reduction of Algarve's rural areas in recent times is escalating and may lead to several insufficiencies in the region. The loss of natural habitats and biodiversity is an increasing concern for regional policy makers and is extensively expressed in the municipal plans of the Algarve.

Methodologically, the key goal of the investigation described in this chapter is to generate similar spatial datasets derived from land use maps and about urban land use. Therefore, the results propose a land use accounting methodology involving the study of population density dynamics and urban growth alterations, for similar time frames. This accounting methodology permits to identify the main driver for agricultural land loss, i.e., it ponders whether urban growth may be considered as a significant driver for loss of rural areas, or, alternatively, systemic population decrease in rural areas might be a key driver of agricultural land appropriation resulting in rural land abandonment, a common concern for the European Union. This is a mostly important conclusion to help decision-makers (local, regional, or national) to choose adequate regional policy for a more balanced development of the region. These variations are registered as urban variations and agricultural land appropriation variations and measured together with the population density outline per municipality. The results of this analysis (see Fig. 9.4) confirm and characterize the responsible driver, as well as a hypothetical evaluation of future trends regarding agricultural land use and population density for the Algarve region. The results, when combined with the quantitative support from spatial data, enable a better understanding of the dynamics of sustainable development, considering that urban growth is an inevitable reality, but that the need for sustainable cities must also be taken into account.

The comparison of the realities of both loss of agricultural land derived from land loss inventories and urban growth will allow us to have a comparative analysis using

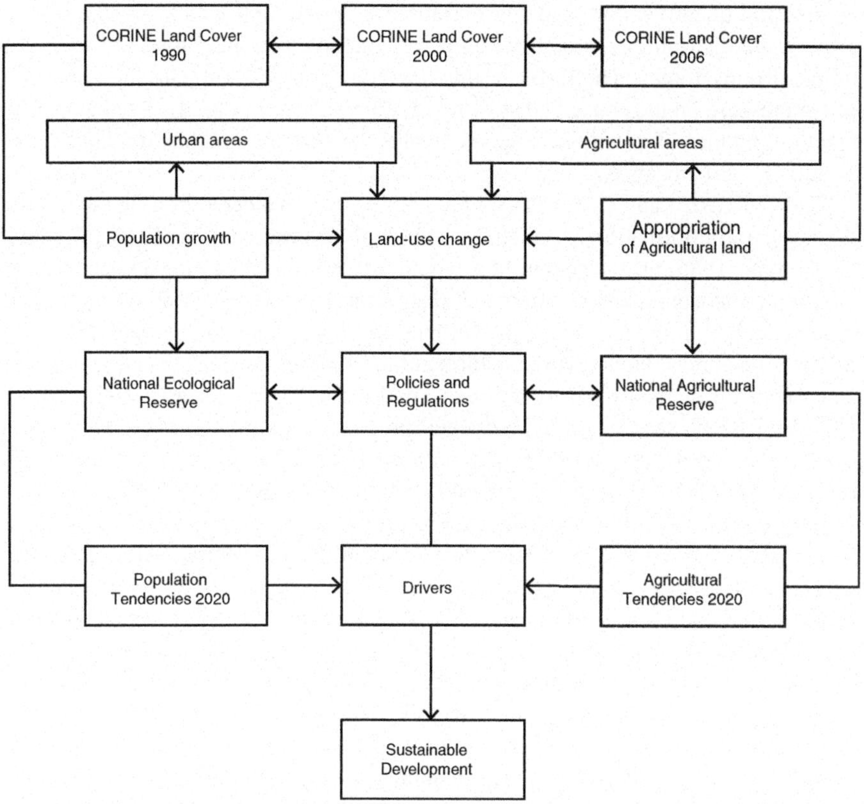

Fig. 9.4 Flow diagram of methodology

spatial information. Figure 9.4 shows the workflow involved in the comparison of the CORINE Land Cover period for 1990, 2000, and 2006 with the population growth surveys conducted from 1991 to 2008. Given the change analysis of the CLC periods, urban areas and agricultural areas are mutually assessed to examine land use change. The diagnosis of population growth and appropriation of agricultural land are reported within the directives of the framework of regulation between 1989 and 2008. The strategies for the National and Ecological Reserve and the National Agricultural Reserve are based on existing policies, designed to sustain the available carrying capacity to take into account the pressure of population tendencies for the period up to 2020 and agricultural trends for the same period. Population increase and urban growth were compared at the three time stamps available for CORINE Land Cover. Population increase showed a predominant tendency to locate at the coastal fringe, while urban sprawl occurred in the same areas where population change was evident. This comparison was made by normalizing population and urban density from 0 to 1, where the normalization of urban density was computed as a result of considering 1, the total urbanization of regions, and 0, the regions with no urbanization. The multiple time series of the population, urban quantification, agricultural quantification, and appropriation of agricultural land, as well as difference in the distribution of density, allow a comparison of land loss and socioeconomic driving forces. The result of this quantification will mitigate the impacts of urban pressure on the coastal areas of the Algarve, as well as recognize the ongoing legal importance of quantitative spatial analysis within the notions of a land appropriation for urbanization.

The CORINE Land Cover (CLC) project may be considered as a first attempt to derive spatial information about land use in the European context. The CLC started on 27 June 1985, as a program that would address the following issues: state of individual environments, geographical distribution and state of natural areas, geographical distribution and abundance of wild fauna and flora, quality and abundance of water resources, land cover structure and the state of the soil, quantities of toxic substances discharged into environments, and list of natural hazards. In this sense, the CLC can be seen as an experimental project for gathering, coordinating, and ensuring the consistency of information on the state of the environment and natural resources in the community (85/338/EEC, Council Decision 27/6/1985).

The Reserva Agrícola Nacional is a Portuguese key instrument for land management, which covers those areas which, due to their promising morphological, climatic, and social conditions, are considered to have the most likely expansion of agricultural activities. Yet, local patterns of agricultural activity, characteristic of the rural areas of the Algarve, have already lost some of their traditional positive externalities, leading to an increase of disadvantageous externalities (negative) created by nonsystemic production sectors. This led to the promulgation of the "Law of Land Use" (Decreto Lei n. 794/76, 5 November) in 1976, which brought policies for urban monitoring and planning of agricultural activity. However, both urban and population pressures as well as the absorption by the secondary and the tertiary sectors promoted further agricultural desertion accompanied by predictable urban growth in Portugal. Based on physical and geographical characteristics (Decreto Lei n. 196/89, 14th June), the RAN is divided into two distinct classes (A and B). As we can conclude from Fig. 9.5, RAN land is systematically decreasing, while urban areas are registering a steady increase.

A closer analysis of land appropriation over the last decades shows a fluctuating pattern especially seen since 1996 (Fig. 9.6). Strongly linked to existing land use policies, the appropriation patterns show an increase since 1994 and in 2005 register the most significant appropriation of RAN land, with a total of 3,722,864 m^2 of land lost.

The evolution of agricultural land appropriation (Fig. 9.5) shows a pattern which is of increasing concern. A closer analysis of urban growth tendencies for the Algarve region, confirms the tendency of agricultural land loss due to urban pressure.

Examination of urban growth patterns between the 1990s and 2006 proves a continuous growth in all the municipalities (Table 9.1).

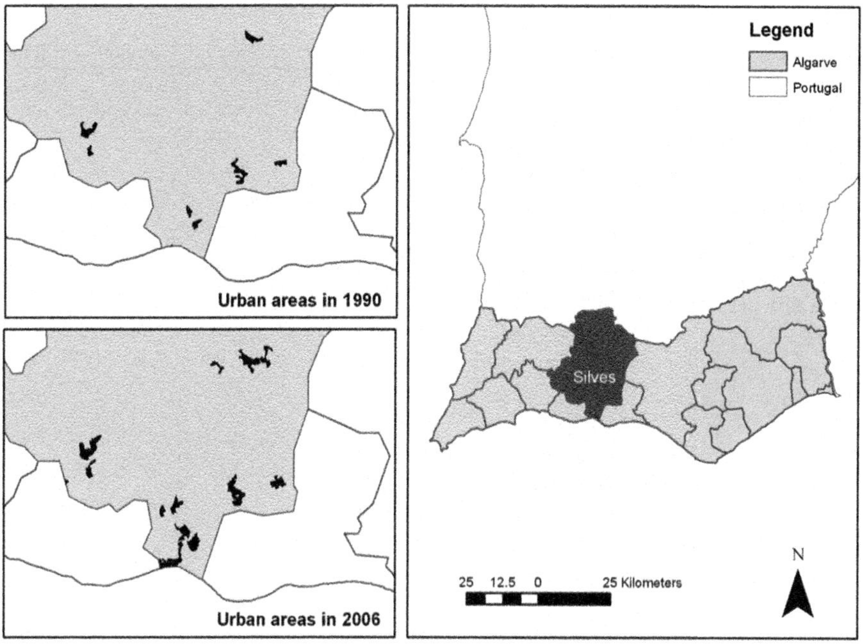

Fig. 9.5 Urban growth change in Silves

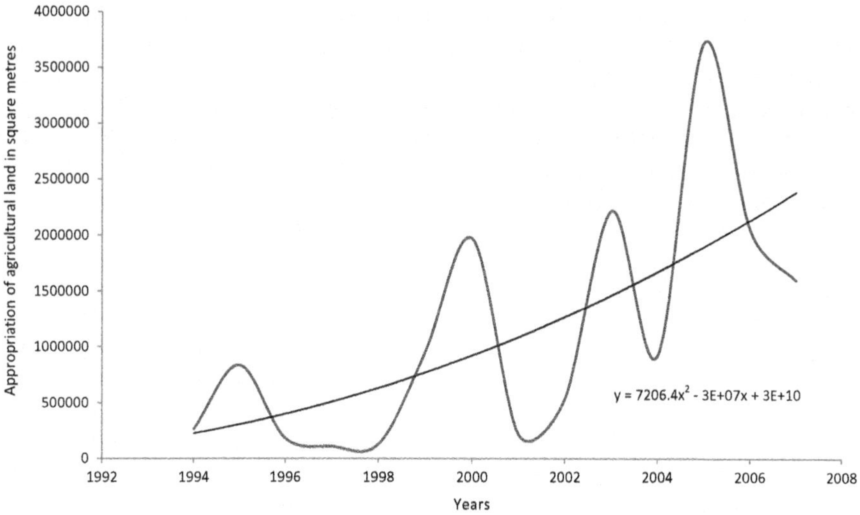

Fig. 9.6 Evolution of agricultural land appropriation

Table 9.1 Ratio of urban land variation from CLC 90 to CLC 06 (in pixels)

Municipality	Urban land 90	Urban land 2006	Variation	Ratio
Castro Marim	3463	14,145	10,682	0.76
Alcoutim	506	2032	1526	0.75
Monchique	1304	5068	3764	0.74
Silves	10,445	25,517	15,072	0.59
Vila do Bispo	6128	14,510	8382	0.58
Tavira	7775	17,942	10,167	0.57
São Brás de Alportel	1836	4149	2313	0.56
Vila Real de Santo António	7816	17374	9558	0.55
Loulé	53,601	116,356	62,755	0.54
Lagos	20,980	44,301	23,321	0.53
Albufeira	30,404	64,128	33,724	0.53
Lagoa	22,264	45,685	23,421	0.51
Olhão	11,407	21,617	10,210	0.47
Portimão	31,276	53,941	22,665	0.42
Faro	21,748	30,750	9002	0.29
Aljezur	10,990	14,446	3456	0.24

Though the municipalities recorded significant increases over the last 15 years of this study, a clear rise of urban sprawl has been confirmed in Castro Marim, Alcoutim, and Silves. It is interesting to note that these areas had a large and traditional agricultural sector having been exposed to a fast urban growth. Additional investigation on the population density patterns in the Algarve, which also reflects such increase, could be a straight effect of competitive prices for building, as well as of existing, road networks which allow communication to important cities such as Faro, Portimão, and Albufeira.

9.4 Conclusions

9.4.1 Urban Growth and Agricultural Land Loss

The comparative analysis of current land use patterns concerning the re-designation of agricultural land for urban use permitted us to relate the results of urban variation per municipality with a variation of losses in RAN. The Algarve has suffered from a severe loss of agricultural areas and a significant reduction of the RAN. Also, the European recent recession enhanced most of the agricultural areas to increasing abandonment in detriment of leapfrogging of peri-urban infrastructure. The double assessment of urban land and changes in the RAN has allowed understanding the impacts of urban concentration in littoral areas and, at the spatial level, perceiving the changing aspects of land use in the Algarve. The municipal information for the Algarve also incorporates the option of understanding local-level impacts of such

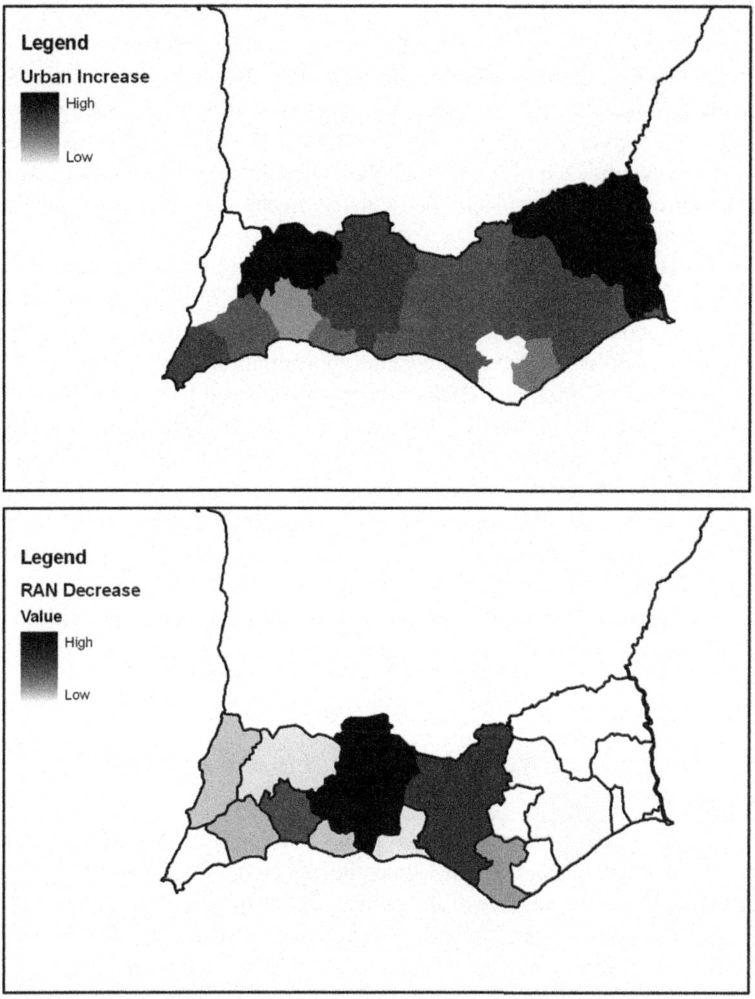

Fig. 9.7 Comparison of RAN decrease and urban increase in the Algarve

changes, such as the case of Silves, where the citric production is traditionally high in parallel to a current urban growth that follows that major trend of construction in peri-urban areas. Datasets of the RAN were summed to the same time frames as CLC data, and information was cross-linked. The results presented Alcoutim, Monchique, and Silves, with urbanist growth, and explicit loss of agricultural land, by overstepping the agricultural regulation of the RAN in the case of Silves. Of course, at the local level, this information is not evident. But a regional quantification of variations of the RAN and urban areas may offer much clarification of land change patterns for the Algarve (Fig. 9.7).

The sum of land use accounting methods benefits from combining different spatial inventories and permits to make better choices and decisions in areas of quick urban and agricultural land use transitions and also due to internal and external environmental, social, and economic pressures. Geographic information systems, thus, should be accepted as tools that allow better hosting local management and supply accurate evidence on existing challenges regarding the landscape as well as the rural European environments, particularly, in the southern and Mediterranean European regions.

While this acknowledgment is obvious, the Algarve shows an increasing tendency of land appropriation, with a rising propensity for urban growth along the appropriated lands. This pattern seems to be combined with a fluctuating tendency of policies in the areas which have allowed the appropriation of agricultural land during certain years (2000, 2003, and 2005), where these appropriations are mainly linked to important regional activities such as the Euro 2004 football championship.

The economic prosperity brought by the tourist industry has stimulated the strategy of the creation of infrastructures within the Algarve. Within the concept that currently promotes the Algarve as a sun and beach district – the "Algarve" – appropriation of agricultural land has led to unavoidable agricultural land loss, especially in peri-urban fringes. The environmental consequences of this growth are evident: traditional agricultural land has greatly decreased, while new infrastructures have increased around certain central areas along the coastal regions.

9.4.2 The Reserva Agrícola Nacional in the Future

On 29 January 2009, a new legislation came to force, Reserva Agrícola Nacional (RAN), under the law 196/1989 on 4 June; the objective of this law is to reinforce the legal nature and the importance of the public administration of the RAN. According to the United Nations definition and nomenclature of territories, methodologically, this classification envisages the better protection of natural resources throughout the country.

For the first time in the regulation history of the RAN, the use of digital information derived from geo-referenced datasets will have an important role in the analysis and synthesis of crucial information for better management. While recently there has been unbalanced management of agricultural land as an inevitable result of the economic growth of urban areas, in the future, the better integration of information could lead to improved decision-making. The Comissão Regional da Reserva Agrícola might have an important role in reshaping the balance of sustainable development for the Algarve.

It is likely that spatial information will have an important role in creating synergy within this commission, allowing more interactive and more soundly based decision-making.

The manufacturer's authorised representative in the EU is Springer Nature Customer Service Centre GmbH, Europaplatz 3, 69115 Heidelberg, Germany. If you have any concerns regarding our products, please contact ProductSafety@springernature.com

Printed and bound by CPI Group (UK) Ltd, Croydon, CR0 4YY

30/09/2026

02225471-0011